Clement Reid, William Whitaker

The water supply of Sussex from underground sources

Clement Reid, William Whitaker

The water supply of Sussex from underground sources

ISBN/EAN: 9783337142506

Printed in Europe, USA, Canada, Australia, Japan

Cover: Foto ©Andreas Hilbeck / pixelio.de

More available books at **www.hansebooks.com**

MEMOIRS OF THE GEOLOGICAL SURVEY.

ENGLAND AND WALES.

THE

WATER SUPPLY OF SUSSEX

FROM UNDERGROUND SOURCES.

BY

WILLIAM WHITAKER, B.A., F.R.S.,

AND

CLEMENT REID, F.L.S., F.G.S.

PUBLISHED BY ORDER OF THE LORDS COMMISSIONERS OF HER MAJESTY'S TREASURY.

LONDON:
PRINTED FOR HER MAJESTY'S STATIONERY OFFICE,
By WYMAN AND SONS, LIMITED, FETTER LANE, E.C.

And to be purchased, either directly or through any Bookseller, from
EYRE AND SPOTTISWOODE, EAST HARDING STREET, FLEET STREET, E.C.,
and 32, ABINGDON STREET, WESTMINSTER, S.W.; or
JOHN MENZIES & Co., 12, HANOVER STREET, EDINBURGH, and
90, WEST NILE STREET, GLASGOW; or
HODGES, FIGGIS, & Co., LIMITED, 104, GRAFTON STREET, DUBLIN.

1899.

Price Three Shillings.

PREFACE.

Ever since its commencement the Geological Survey of the United Kingdom has given much attention to the question of water supply, and has accumulated a large body of information on the subject relative to all parts of the country. It has thus been able constantly to give assistance to professional men and others who have sought for advice in regard to the geological problems involved. In order to make the practical work of the Survey in this department more useful to the general public, it is now proposed to issue a series of Memoirs which, dealing with the underground waters of the different counties, may aid local effort in obtaining supplies of wholesome, uncontaminated water. In these Memoirs only such geological details will be given as may bear directly upon the question of water supply. The present account of the underground waters of Sussex is the first of the proposed series.

While the Geological Survey has been ready to furnish geological information, it has, in turn, received much assistance from those practically concerned in dealing with water supply. Engineers and well-sinkers have willingly lent their aid, without which it would have been impossible to gather the numerous facts of which the Survey is now in possession. In the preparation of the present Memoir, for example, we have been specially indebted to Mr. P. H. Palmer, the Borough Engineer of Hastings, and to Mr. E. Easton; also to the communications of the well-sinkers Messrs. Duke and Ockenden, Messrs Le Grand and Sutcliff, Messrs. Docwra, and Messrs. Isler and Co. The contributions of these collaborators, as well as those from other sources, are acknowledged in each case. About 150 of the records of wells in the following pages are now published for the first time.

The late Mr. W. Topley took part in the collection of materials for the present Memoir, but his much-lamented death has deprived us of remarks on many of the sections which he could have supplied better than anyone else.

In issuing this first publication on the water supply of the English counties, I desire to point out to those interested in the subject that it would be greatly for the public benefit if full details of all the strata passed through in sinking wells were in each case forwarded to the Geological Survey Office in order that they might be placed among the permanent records of water supply. The name of each informant would, of course, be given when the information supplied by him was published. It would be desirable, at the same time, to have information regarding the yield of water, and to obtain copies of any chemical analyses that might be made. Many of the records supplied to us are less valuable than they would be for want of information as to these particulars.

The detailed geological structure of the County of Sussex is given in the Maps and Memoirs of the Geological Survey enumerated on p. 7 of the present publication. For general purposes of reference the two sheets (12 and 15) of the Index Map of England and Wales, on the scale of four miles to one inch, will be found useful. Fuller information is given on the maps on the one-inch scale, while, where further local details are specially required, they can be obtained from MS. maps on the scale of six inches to a mile, which can be supplied at the cost of transcription for those parts of the county which have been revised and published in the new series of one-inch maps.

It should be added that Mr. Whitaker, though he has retired from the Survey, has been good enough to find time, during his tenure of the office of President of the Geological Society, to assist in arranging and connotating the well-sections in the present Memoir and in correcting the proofs.

ARCH. GEIKIE, Director-General.

Geological Survey Office,
    28, Jermyn Street, London.
    31st October, 1898.

CONTENTS.

	PAGE.
Preface by the Director-General	iii
Introduction: Outline of the Geology as far as relates to Water Supply. Mean Annual Rainfall. List of Geological Survey Works on Sussex	1
Well Sections in Sussex	8
Analyses of Waters	103
Index	122

THE WATER SUPPLY OF SUSSEX

FROM UNDERGROUND SOURCES.

INTRODUCTION.

Sussex, for various reasons, is largely dependent on deep wells for its water-supply. Good surface-springs are comparatively rare, the running streams soon become turbid, and shallow wells in loose superficial deposits are so liable to contamination that the increasing population renders them every year more unsafe. As the amount and quality of the water to be obtained from a deep well depends mainly on geological considerations, a short account of the geology is prefixed to these records, to help those who desire to obtain water at new localities.

OUTLINE OF THE GEOLOGY AS FAR AS RELATES TO WATER SUPPLY.

The upheaval of the Weald, which caused nearly all the streams to flow north or south, away from the central axis, still causes the underground waters over the greater part of the county to flow southward, in the direction of the dip. The upward arching of the strata, and the subsequent removal by denudation of the higher parts, have brought within reach so many different formations that we must here deal with the whole of the following series :—

		Character of the water in Sussex.
Recent	Blown Sand	usually bad, salt, and supply small.
	Shingle	
	Alluvium	very bad.
Pleistocene	Brickearth	none.
	Valley Gravel	fair, somewhat ferruginous, but very liable to surface contamination.
	Coombe Rock	very hard; liable to surface contamination.
	Raised Beach	water has percolated through Coombe Rock, and is of similar character.
	Plateau Gravel	variable, small quantity, and liable to contamination.
	Clay with Flints	none.
Eocene	Bracklesham Beds	generally bad.
	Bagshot Beds	probably ferruginous, and quantity small.
	London Clay	none.
	Woolwich and Reading Beds	a little in the sandy beds, generally containing iron and sulphates; none elsewhere.

Upper Cretaceous	Upper Chalk - - -	}	abundant supply of good water, hard with carbonates (temporary hardness), less so with sulphates (permanent hardness).
	Middle Chalk - - -		
	Lower Chalk - - -		a little hard water, usually with excess of sulphates.
	Upper Greensand - -	-	good, comparatively soft.
	Gault - - - -	-	none.
Lower Cretaceous	Folkestone Beds - -	-	slightly ferruginous, but good.
	Sandgate Beds - -	-	little.
	Hythe Beds - - -	-	much, good quality.
	Atherfield Clay - -	-	none.
	Weald Clay - - -		occasionally a little in sand beds.
	Tunbridge Wells Sands	-	good, quantity uncertain.
	Wadhurst Clay - -	-	water in the rock-beds.
	Ashdown Sand - -	-	good, slightly chalybeate.
	Fairlight Clay - -	-	none.
Upper Jurassic	Purbeck Beds - -	-	probably none.
	Portland Beds - -	-	perhaps some.
	Kimeridge Clay - -	-	none.
Middle Jurassic	Corallian - - -	-	little or none.
	Oxford Clay - -	-	none.

Water-supply being the sole question to be dealt with in this Memoir, no geological details are given which do not refer directly to the water-bearing or retentive character of the rocks, the quality of the water, or to the probable continuity of the various strata, on which last depends the amount of water which we may expect to find. Those wishing to study the geology from other points of view will find further particulars in the Geological Survey Memoirs which deal with the various parts of the county, or in Dixon's "Geology of Sussex" (2nd ed., 4to, Brighton, 1878).

Palæozoic Rocks.

The prospect of obtaining water from the Palæozoic Rocks under any part of Sussex is so slight that we could not advise the sinking of any trial-bores. In the first place, the least depth of these rocks from the surface is probably fully 2,000 feet, and any water found at that depth would be distinctly warm. Secondly, over the whole county, and far beyond its limits, thick masses of impervious clays occur, so that the only sources of supply would be from distant areas, where Palæozoic rocks crop out, or are overlaid by porous strata. Thirdly, rocks at such depths are so compressed by the weight of overlying strata that they seldom contain open fissures or yield much water.[*] The deep borings and sinkings in search of coal now being made in the adjoining parts of Kent may, however, cause us to modify this opinion.

---

[*] See also Prestwich, "Proc. Inst. Civ. Eng.," vol. xxxvii., p. 126.

Jurassic Rocks.

The oldest rocks yet met with in the county are those penetrated by the "Sub-Wealden Boring" at Mountfield, near Battle. In this boring the Oxford Clay, Corallian Rocks, and Kimeridge Clay were either shales or very shaly, and yielded no water. The Portland Beds consisted mainly of sandstone. When the tools penetrated the Purbeck rocks at 169 feet, the column of water in the bore-hole sank 40 feet, thus proving that the Portland Beds are pervious, and at lower levels may yield water. The quality of the water, however, would be uncertain, for if the springs are fed by percolation through the gypsum-bearing overlying rocks there would probably be an excess of sulphates, and the quantity also would not be large. There is also a possibility that rocks in such a position might yield natural gas or petroleum instead of water. In the Purbeck Beds no water is found, except small springs highly charged with sulphates.

Fairlight Clay.

In the neighbourhood of Hastings occurs a mass of clay over 300 feet in thickness, with subordinate beds of sand. Its exact relation to the Purbeck and Wealden strata is still in doubt. No water is found in this clay.

Ashdown Sand.

This sand is found at the surface over a large area in the middle of the Weald. It contains water of good quality, but like most of the Wealden sands is too fine-grained usually to yield any large supply from a boring. It should be noticed that sunk wells may succeed where borings fail, for the amount yielded by such strata depends largely on the surface exposed in the well. The Ashdown Sand often contains beds of clay, which must greatly hinder the circulation of the water. Hastings is supplied from this source.

Wadhurst Clay.

The Wadhurst Clay is 130 feet thick or more. It yields no water, but over the area occupied by it a moderate supply can usually be obtained by boring to the Ashdown Sand below.

Tunbridge Wells Sand.

This sand varies from 160 feet in the east to 380 feet near Cuckfield, two masses of clay, the Grinstead Clay and the Cuckfield Clay, coming in where the deposit is thickest. Mr. Topley's estimate at Cuckfield gives :—

		Ft.
Upper Tunbridge Wells Sand.	{ Sand and sandstone, with layers of Tilgate stone at the top	115
	Cuckfield Clay	15
	Sand and sandstone	70
Grinstead Clay	Clay and shale	80
Lower Tunbridge Wells Sand.	} Sand and sandstone	100

The water from the sand is good, but commonly ferruginous. It is doubtful whether water can travel freely through it for any great distance, and except where bare or covered by a small thickness of Weald Clay it cannot be depended on. Borings reaching this sand below any great thickness of clay may yield enough for isolated houses or small villages, but will seldom supply a large quantity.

Weald Clay.

The Weald Clay, though usually quite devoid of water contains occasional beds of sand, which sometimes yield good water in unexpected places. These sands are impersistent and their position cannot be forecast with any certainty. The supply to be expected from them is probably nowhere large. The clay is probably over 600 feet thick at the northern border of the county, but appears to diminish considerably in thickness towards the east. It has, however, been proved to a depth of 200 feet at Eastbourne, without reaching the base.

Lower Greensand.

These sands are usually of coarser grain than those of the Wealden Series, and consequently yield their water more freely, and are less liable to choke the bore-hole. The water is either soft and very pure, or else somewhat ferruginous, or it may contain sulphates; it does not usually show excess of lime. Between Eastbourne and Lewes this formation has become very thin, or is only represented by the highest division, and cannot be depended on for anything but a small supply. West of Lewes it thickens rapidly and can be divided into four series :—

Folkestone Beds : coarse sands, 12 to 140 feet.
Sandgate Beds : fine-grained sand and clay, 30 to 100 feet.
Hythe Beds : sand, sandstone, and chert, calcareous above, 25 to 200 feet.
Atherfield Clay : shelly clay, without water.

The Hythe Beds and the Folkestone Beds can be depended on to yield a fair supply in most localities. The Sandgate Beds are uncertain, except sometimes in shallow wells and near the outcrop.

Gault.

The Gault in Sussex is about 300 feet thick, and is always impervious and without water. It is often advisable, however, to commence a well in the Gault, for water obtained from the the Lower Greensand in this way is free from any suspicion of direct surface-contamination.

Upper Greensand.

The Upper Greensand is a glauconitic sand or sandstone, calcareous in the upper part, and from 40 to 80 feet in thickness. The general narrowness of its outcrop in Sussex makes it difficult to obtain a large supply from this source. The water, however, is moderately soft and of excellent quality. Headings in the rock-bed of the Upper Greensand till lately supplied Eastbourne. The water, however, is apparently to a large extent derived from the Chalk above, and is let in by the exceptionally fissured state of the rocks in that district.

Lower Chalk.

This division consists of from 160 to 200 feet of alternating well-bedded grey chalk and chalk marl with pyrites. It is too impervious to yield water, except where so shattered that fissures let in water from the more pervious chalk above Springs originating in this way are seen on the foreshore at Holywell, and at Jevington, near Eastbourne. Small springs from the Lower Chalk yield water with excess of sulphates.

Middle Chalk.

The Middle Chalk includes about 200 feet of hard rubbly chalk, with a few flints in the upper part. Towards the base, where it rests on the impervious grey "Belemnite Marl," which forms the top of the Lower Chalk, occurs about 10 feet of hard rock, the Melbourn Rock, from which are given out many springs, like those in the cliff at Holywell. The hardness of this rock, and the consequent openness of the fissures, makes it advisable to continue borings to this level, in cases where the Chalk above has proved too compact to let in water.

Upper Chalk.

This division is about 700 feet thick in West Sussex, but thins to 500 feet at the east end of the South Downs, through the loss of the upper part before the Eocene strata were deposited. It consists of soft chalk with flints. The water from the Middle and Upper Chalk is hard, but can readily be softened.

Woolwich and Reading Beds.

These are principally clays, and where sandy are full of pyrites; they cannot be recommended as a source of water-supply

London Clay.

The London Clay is about 300 feet thick in Sussex. It contains a few beds of loamy sand, especially in the uppermost and lowermost parts, and at the base is sometimes found a mass of flint-pebbles. These have nowhere yielded a supply, though a little ferruginous water is sometimes met with.

Bagshot and Bracklesham Beds.

The Bagshot Sands in Sussex are thin and inseparable from the Bracklesham Series. These latter are apparently 500 or 600 feet thick near Selsey. Occasionally drinkable water is found in them, but usually the supply is small and the taste very unpleasant.

Drift Deposits.

The character of the water to be found in these is sufficiently set forth in the table. In all cases there is risk of contamination in shallow wells, though isolated farms and houses are perforce obliged to put up with water from this source. The usual situation of these wells, surrounded by farm-buildings, is particularly objectionable; a site in the middle of a lawn or garden is preferable, and greater care should be taken to place any cesspools as far as possible from the well.

MEAN ANNUAL RAINFALL OF SUSSEX.
(From "Rainfall Tables of the British Islands, 1866–1890.")

	Height above Mean Sea Level.	Period of Observation.	Mean Rainfall.
Arundel, Dale Park	316	1866–80	34·29
Balcombe Place	300	1866–80	34·17
Brighton	55	1881–90	28·33
Chichester, Westgate	40	1866–80	29·16
Chilgrove, near Chichester	284	1866–90	33·74
Crowborough Beacon	777	1871–90	36·81
Cuckfield, Borde Hill	270	1881–90	29·73
Eastbourne	15	1871–90	31·60
East Grinstead	365	1866–90	32·72
Fernhurst (Haslemere)	301	1866–80	32·19
Glynde Place, near Lewes	49	1866–90	32·60
Hastings, Hollington	320	1866–90	29·19
Littlehampton	20	1881–90	27·48
Midhurst, Lynch	160	1866–80	39·65
Petworth Rectory	180	1866–90	34·75
St. Leonards	130	1881–90	29·08
Uckfield Observatory	149	1866–80	31·02
Uckfield	200	1881–90	29·42

It may be observed that there are no records from the higher parts of the South Downs, on which occurs, apparently, the heaviest rainfall. These Downs are the first hills to intercept the moist air from the south-west. On the Downs above four hundred feet the condensation of mist also is considerable in the autumn and winter, often causing moisture to drop from every leaf, though in the towns below it is quite dry. This condensation supplies the dew-ponds.

List of Geological Survey Works on Sussex.

Sheets of the Index Map. Scale four miles to one inch.

12. Northern half of the county.
15. Southern half of the county.

Sheets of the Map. Old Series. Scale an inch to a mile.

4. Western part. Rye. By F. DREW. 1863.
5. All but the north-eastern part. Battle, Eastbourne, Hastings, Lewes, Seaford, Winchelsea, and Ashdown Forest. 1864. By W. T. AVELINE, H. W. BRISTOW, F. DREW, C. GOULD, T. R. POLWHELE, C. LE N. FOSTER, W. TOPLEY, and W. B. DAWKINS. Chalk-divisions and Drift over the Chalk-tract added 1893. By W. A. E. USSHER and C. REID.
6. Strip on the south (western and central parts). East Grinstead. 1864. By F. DREW. Drift Edition, 1886. (Little Drift in the Sussex part.)
8. Strip on the south (eastern and central parts). 1862. By F. DREW. Drift Edition, 1887. (Hardly any Drift in the Sussex part.)
9. All but a narrow strip on the north (western part). Arundel, Bognor, Bramber, Brighton, Chichester, Cuckfield, Horsham, Littlehampton, Midhurst, Petworth, Shoreham, Steyning, Worthing, and Selsea Bill. 1864. By H. W. BRISTOW, F. DREW, C. GOULD, J. HAY, F. C. BISHOPP, and W. B. DAWKINS. Chalk-divisions and Drift over the Chalk-tract added 1893. By C. REID.

Sheets of the Map. New Series. Scale an inch to a mile.

331. North-eastern corner. 1893.
332. Bognor, Littlehampton, and Selsea Bill. 1893.
333. Worthing (part). 1893.
334. Seaford, Eastbourne. 1893.

Sheets of the Horizontal Sections. Scale six inches to a mile.

73 (part). From Selsea Bill to Siddlesham, Chichester, East Lavant, Singleton, Cocking, Midhurst, and Haslemere. 1868.
75 (part). From W. of Worthing to Cisbury, Chanctonbury, Shipley, Itchingfield, and near Horsham. 1867.
76 (part). From E. of Kemp Town, Brighton, to Warren Farm (Brighton Industrial Schools), Stanmer, Ditchling Beacon, Wivelsfield, Haywards Heath, Wakehurst Park, and Rowfant. 1867.
77 (part). From W. of Newhaven Harbour to Piddinghoe, Mount Caburn (near Lewes), Little Horsted, near Uckfield, to Buxted, across Ashdown Forest to Crowborough Beacon, and near Groombridge. 1867.
78 (part). From Beachy Head, across the Downs, to Polegate, Hailsham, near Heathfield, Ticehurst Road Station, and Ticehurst. 1867.

Memoirs, 8vo.

The Geology of the Weald (parts of the Counties of Kent, Surrey, Sussex, and Hants). By WILLIAM TOPLEY. 1875.

The Jurassic Rocks of Britain, Vol. V. The Middle and Upper Oolitic Rocks of England (Yorkshire excepted). By H. B. WOODWARD. 1895.

The Geology of the country around Bognor. (Explanation of Sheet 332.) By CLEMENT REID. 1897.

The Geology of the country around Eastbourne. (Explanation of Sheet 334.) By CLEMENT REID. 1898.

WELL SECTIONS IN SUSSEX.

[Words, etc., in square brackets have been added by us.]

ALDINGBOURNE. Headhone Farm.

Blue [London] Clay - - - - - - 235 ⎫ 335 feet.
[Reading Beds.] Red mottled clay - - - 100 ⎭

[Must have stopped within 10 feet of the Chalk.]

ANGMERING.

Communicated by Mr. R. WINCHESTER, 1896.

		Thickness.	Depth.
		Feet.	Feet.
[Drift]	Clayey gravel, "shrave"	12	—
	Clean loam	4	16
	Sand	10	26
[Upper Chalk]	Chalky marl, with water	12	38

ARUNDEL. Coal Yard.

[Recent Deposits] { Marsh clay - - - - 20 ⎫ 38 feet.
                   { Sand and shingle - - - 18 ⎭
[Upper Chalk].   Marl and chalk.

ARUNDEL. Mr. T. Barnes'.

From MR. CRAWFORD.

Water-level 30½ feet down.

		Thickness	Depth.
		Feet.	Feet.
Dug well, the rest bored		—	8
[Reading Beds]	Clay and freestone	29	37
	Mottled clay and sand	4	41
	Black sand	5	46

ASHBURNHAM Place.

W. TOPLEY, "Geology of the Weald," p. 65, 1875.

Wadhurst Clay, with a few inches of rock, 62 feet.

In a shallower well, at a cottage north of the Parsonage, there were a few thin beds of sandstone full of fossils.

BALCOMBE. Mid-Sussex Waterworks. 1890.
Communicated by MR. J. CHURCH.
Shaft of 8½ feet diameter. Water level 248¼ feet down.

Compared with springs in the neighbourhood the water is very free from iron. The pumps in use for sinking could only lift 120 gallons a minute (= 172,800 a day of 24 hours), and could not keep the water down (after seven days' pumping), **therefore the work was** suspended. The yield has been 180,000 gallons in 24 hours.

		Thickness.	Depth.
		Feet.	*Feet.*
[Tunbridge Wells Sand]	Yellow sand, with bands of ironstone	13	13
	Rock	5	18
	Sand and clays	32	50
	Spring tapped here [bed not described]	1	51
[Tunbridge Wells Sand or Wadhurst Clay, 47 feet]	Hard grey clay	20	71
	Coloured [mottled] clay	12	83
	Yellow clay and sandstone	9	92
	White sandstone	6	98
	Clay	2	100
[Wadhurst Clay, 109½ feet]	Clay and rock, with water	24	124
	Blue sandy clay, much water	8	132
	Bluish-green clay, giving off carbonic acid	2	134
	Light-blue sandy clay	4	138
	Hard grey clay and shale	24	162
	Very dark shales and clay	45½	207½
	Hard white and grey sandstone	45	252½

The following pumping-tests have been made :—
October 25th and 26th, 1897.
Total water pumped in 5 hours 14 minutes (at intervals during three days), 105,000 gallons.
November 1st, 1897.
Total pumped in 11 hours 25 minutes, 98,082 gallons.
December 6th and 7th, 1897.
On the 6th, yield in 12 hours 90,585 gallons.
On the 7th,   ,,   5   ,,   38,350   ,,
For Analyses of the water, see p. 104.

BARCOMBE. Sewell's Farm, over a mile N.N.W. from the Church. 1883.
Made and communicated by MR. G. BATES, of Lewes. Good supply.

	Thickness.	Depth.
	Feet.	*Feet.*
Brown mixed earth	30	30
Light-blue clay	60	90
Blue clay and mud Sand and hard veins of slate rock	60	150

BARNHAM. Half a mile E.S.E. of the Railway Station.

Sunk and communicated (from memory) by MR. OCKENDEN, SEN.

To Chalk [Drift, London Clay, and Reading Beds] ... 185 } 206 feet.
Chalk. Good spring ... . 21 }

The Station is in EASTERGATE (which see).

BATTLE. The Brewery.

W. TOPLEY, "Geology of the Weald," p. 65, 1875.

Wadhurst Clay. Shale, with a bed of Tilgate stone, 2 feet thick, 40 feet down, 60 feet. Similar stone met in other wells here.

BATTLE. Waterworks N. of the town. 1890 (second boring).

Communicated by MR. J. CHURCH.

Most water comes in on the western and south-western sides of the well, chiefly from the rock below 144 feet.

Where brickwork occurs the beds are soft and shaly (except in old well).

		Thickness.	Depth.
		Ft. in.	Ft. in.
	Old Well	—	35 0
	Rock	2 4	37 4
	Brickwork	2 6	39 10
	Rock	3 4	43 2
	Brickwork	16 1	59 3
	Rock	2 10	62 1
[All Ashdown Sand; or probably the lower part Fairlight Clay, more sandy in this direction.]	Brickwork	11 4	73 5
	Rock	2 0	75 5
	Brickwork	4 3	79 8
	Rock	9 7	89 3
	Brickwork	26 8	115 11
	Rock. Water	2 6	118 5
	Brickwork	10 0	128 5
	Rock	10 2	138 7
	Brickwork	6 0	144 7
	Rock (hard sandstone, W.T.). Water	8	152 9
	Brickwork	5 0	157 9
	Hard Rock, Ironstone	1 6	159 3

MR. TOPLEY has left the following short notes of wells at Battle:—

1. On the eastern side of Mount Street, between a quarter and a third of a mile from the Abbey Gate-house, 60 feet to water.

2. At the workhouse, on the northern side of Northrade Road, nearly a mile west of the town, 65 feet deep, not much water.

3. At North Lodge, east of the workhouse, 70 to 80 feet deep, very little water.

4. In the field south of the road a quarter of a mile south-west of Parkdale (nearly a mile south-west of the railway station), 35 feet deep. Gravel 6 feet and then sand 26.

5. Telham Farm, nearly 1½ miles south-east of the church, 149 feet deep and mostly in rock.

BATTLE (Sub-Wealden Exploration), see MOUNTFIELD.

BEDDINGHAM. Courthouse Farm (near the Church).

Boring. Samples, &c., communicated by MR. KILLICK.

About 25 feet above Ordnance Datum.

Water overflows. Good supply of soft water (for Analysis, see p. 105).

		Thickness.	Depth.
		Feet.	*Feet.*
	Made ground	2	2
Upper Greensand.	Very fine-grained marly green sand	18	20
	Clayey greensand	10	30
	Grey very sandy clay	10	40
	Dark-grey sandy clay (samples at 40, 50, and 60 feet)	30	70
Gault, 310 feet.	Dark soapy clay (samples at 79 and 100 feet)	130	200
	Dark soapy clay and fossils (samples at 200, 283, 300, and 336 feet)	138	338
	Clayey greensand (water at base)	2	340
Lower Greensand.	Loose, very green coarse sand, full of water	6	346

BEDDINGHAM. Toy Farm, about 2½ miles south-east of the Church. Well, 1893.

Communicated by MR. T. W. PICKARD.

About 250 feet above Ordnance Datum.

Average height of water 5 feet. Average yield 300 gallons a day.

White Chalk with veins of flints, 124 feet.

BEEDINGWOOD [? Lower Beeding], near Horsham. Stone Lodge.

Communicated by MESSRS. G. ISLER & Co.

Water-level 93 feet down.

	Thickness.	Depth.
	Feet.	*Feet.*
Shaft (the rest bored)	—	87
Clay	30	117
Clay and rock	3	120
Rock	5½	125½

BEXHILL, 1851.

Communicated by Messrs. Docwra.

	Thickness.	Depth.
	Feet.	Feet.
Clay	19	19
Stone	5	24
Clay	1	25
Stone	10	35
Hard dead sand	30	65
Petrified wood	$\frac{1}{2}$	$65\frac{1}{2}$
Coloured clay	$4\frac{1}{2}$	70
Sand rock	8	78
Coloured clay	6	84
Stone	1	85
Claystone	2	87
Coloured clay	30	117
Dead black sand	8	125
Boggy stuff	2	127
Coloured clay	16	143
Green sand	1	144
Rock	$1\frac{1}{2}$	$145\frac{1}{2}$
Dark dead sand	$11\frac{1}{2}$	157
Dark clay	23	180

BEXHILL, for Mr. G. Lane. 1851.

Sunk and communicated by Messrs. Docwra.
Water of only 1 degree of hardness rose to 48 feet from the surface.

	Thickness.	Depth.
	Feet.	Feet.
Light[-coloured] sandy clay	30	30
Blue clay	76	106
Blue shaly rock	24	130
Coloured clays [5 beds]	80	210
Grey sand	30	240

BEXHILL. Mr. J. C. Kenwood's.

Boring. Made and communicated by Messrs. G. Isler & Co.
Water-level 37 feet down. Supply abundant.

	Thickness.	Depth.
	Feet.	Feet.
Sandy clay	15	15
Blue clay	8	23
Soft sandstone	28	51
Hard blue clay	9	60

BEXHILL. Waterworks. Boring, in the Marsh, less than half a mile S.E. of Buckholt Farm. 1892.
Communicated by MR. W. B. LEWIS.

Twelve feet above Ordnance Datum. Water good. September 1893. Pumped day and night 300,000 gallons. A letter from MR. LEWIS (October 1894) adds that when about 260,000 gallons are pumped in 24 hours it about balances the ordinary flow.

		Thickness.	Depth.
		Feet.	*Feet.*
Sump		—	25
[Mapped as Ash-	Marl	13½	38½
down Sand, but	Clay	71	109½
apparently only	Blue stone, the bottom foot hard	5½	115
so in part]	Marl and clay	20	135
	Blue sand-rock	18	153

MR. W B. LEWIS's letter of 1894 says that some recent boring, close to the sump, does not encourage the belief that more water would be got by deepening; and that a boring near Sidley Brook, made in 1891, gave no promise of water in sufficient quantity.

A note by MR. TOPLEY says that the old well [? Wrest Wood] is 114 feet above Ordnance Datum, and is a shaft of 122 feet. The water was pumped out in 3½ hours, at the rate of 6,000 gallons an hour; but 45,000 to 50,000 gallons a day can be got. The water comes into the heading more at high tide than at low. The heading being 22 feet both eastward and westward from the shaft.

The pumping from this well has drained the wells at Buckholt Farm and at Henniker Farm (less than half a mile W.S.W. from the works). Note by MR. TOPLEY.

BEXHILL. Waterworks. New Well (? first) in the Valley.

The following notes by MR. TOPLEY (1890) may refer to the well above described, but they differ much from the description given.

Well 18 feet above Ordnance Datum.

Sump 12 feet square and 25 feet deep, then bored.

Water rose to 4 feet from the surface, from the more open sandstone. At first 40,000 gallons a day got from the top sandstone, then this fell to 24,000, and then to 15,000.

	Thickness.	Depth.
	Feet.	*Feet.*
Soil, passing into sandstone, some hard, mostly broken, with partings	38	38
Blue clay	57	95
Blue marl and hard layers	10	105

Another note of MR. TOPLEY's mentions a New Well, N.W. of Crouch Farm [? site], as passing through the following beds:—

Soil ...	1 foot
Sandy marl and clay ...	32 feet
Blue clay.	

According to another note (November 1893), a borehole, 50 or 60 feet, a little S. from the new well, gave this section:—

	Thickness.	Depth.
	Feet.	Feet.
Peat	8	8
Yellow sand	22	30
Blue clay	6½	36½
Blue sand	9½	46
Blue and brown sand	5	51
Clay and stone (hard and shelly)	(?) 22	(?) 73

The water in the new well stood 18 feet down when pumped, but rose to the surface when pumping ceased; 250,000 gallons pumped per day of 24 hours (106,000 for Hastings). When 360,000 were pumped air was drawn in.

For another well for Bexhill Waterworks, *see* p. 101.

BEXHILL. Trial for Coal, 1804–1809 (near the Shore).

DR. MANTELL. "The Fossils of the South Downs . . ." 4to. London, 1822, pp. 35, 36.

Shaft 27 feet, the rest bored.

	Thickness.		Depth.	
	Ft.	in.	Ft.	in.
Soil, clay and sandy loam	9	0	9	0
Dark clunch	9	0	18	0
White rock with kind partings	13	0	31	0
Dark clunch	3	0	34	0
Grey rock	5	0	39	0
Dark clunch	3	0	42	0
Strong grey rock	5	6	47	6
Blue binds	3	6	51	0
Grey rock with kind partings	18	0	69	0
Blue bind	3	6	72	6
Stone grey rock	3	0	75	6
Blue bind	2	7	78	1
Strong white rock	4	4	82	5
Dark clunch	7	9	90	2
Smut coal	2	3	92	5
Grey bind	14	3	106	8
Blue bind with iron-ore	10	9	117	5
White stone	3	0	120	5
Clunch or fire-clay	3	2	123	7
White sandstone	5	9	129	4
Kind clunch parting	0	8	130	0
Brown sandy rock	2	9	132	9
Sharp peldron	9	0	141	9
Blue bind	5	0	146	9
Strong brown rock	4	0	150	9
Blue bind, with impressions of fern-leaves	7	6	158	3
Blue bind with iron-ore	2	0	160	3
Strong coal	3	6	163	9

Mr. Topley has remarked of this section. "Some seams of lignite were passed through, reported to vary in thickness from 2 feet 3 inches to 4 feet 6 inches [should be 3 feet 6 inches]; the thickest seam is said to be of bad quality and very sulphureous. These seams are thicker than any known to occur on the surface; and supposing the section to be reliable, it is very remarkable that the shaft should happen to be sunk at a spot where these beds, usually thin and very inconstant, had attained their greatest known thickness. It is, however, very doubtful if these beds really were found, or there would surely have been some more serious attempt to work them. Lower speaks of sanguine adventurers being induced to sink a shaft here, and he adds "adventurers of another kind encouraged the scheme, and fictitious specimens of coal were brought to the surface."* ("Geology of the Weald," p. 348.)

There being some local interest in the matter, it seems well to reproduce the above details, although partly in terms not used for these southern beds.

Birdham. Holt Place.
Communicated (from memory) by Mr. Ockenden, Senr.

Loam	15	} 250 feet.
London Clay	235	

Bognor. Waterworks. (See also Eastergate and Merston.)
From a lithographed section, communicated by Mr. J. W. Grover, C.E. (published as a woodcut in *The Builder*, 25th March, 1876).

Shaft and cylinders 80 feet, the rest bored.

Water-level, without pumping, 20 feet down, giving 150,000 gallons a day nearly 80 feet down.

		Thickness.	Depth.
		Feet.	*Feet.*
	Brickearth - about	9	9
[Drift, 24 feet]	Running sand, saturated with water, which supplied the town about	15	24
[? London Clay]	Red and blue clay - ,,	34	58
	(Undescribed bed) - ,,	4	62
	Wet sand - ,,	5	67
[Reading Beds]	Red and blue clay - ,,	47	114
	Marl rock [may be top of chalk] - about	4	118
Chalk, with flints at 120, 170, and 190 feet down		212	330

Bosham. At the Gatehouse a quarter of a mile E. of the Station.
Bored and communicated by Messrs. Duke & Ockenden.

[Reading Beds] Mottled clays	88	} 150 feet.
Chalk, very soft, with good water	62	

Bosham Harbour. The Duke of Gloucester.
Bored and communicated by Messrs. Duke & Ockenden.

Brickearth, &c. *	18	} 45 feet.
Chalk, very soft	27	

Plenty of good water, but the well cannot be kept clear.

---

* "History of Sussex," vol. i., p. 49.

BRIGHTON. North Street. Messrs. Smithers' Brewery. 1889.
Boring made and communicated by MESSRS. LEGRAND & SUTCLIFF.
Water level 100 feet down.

Old dug well (the rest bored) - - - - - 102 ⎫
Hard chalk and flints - - - - - - 50 ⎬ 152 feet.

BRIGHTON. Waterloo Street. Messrs. Robins' Brewery. 1885.
Made and communicated by MESSRS. LE GRAND & SUTCLIFF.
Water level 28 feet down

		Thickness.	Depth.
		Feet.	*Feet.*
[Drift]	Dug pit (the rest bored) -	---	$10\frac{1}{2}$
	Clay and flints -	4	$14\frac{1}{2}$
	Sand - - -	$\frac{1}{2}$	15
	Chalk and flints -	50	65

BRIGHTON. Waterworks. Two Pumping Stations.
For a third see PATCHAM. For Analyses, see pp. 105, 106.

These works are one of our best examples of a large supply from the Chalk. They have been described in the following papers, from which particulars have been taken, supplemented by information from Mr. J. JOHNSTON, the present engineer :—

1882. E. EASTON. Transactions of the Brighton Health Congress, 1881, pp. 48-56, three plates. Separately printed, 16 pp., 8vo.

1886. W. WHITAKER. *Geol. Mag.*, dec. iii., vol. iii., pp. 159-161. Reprinted in *Public Health* some years later.

1890. W. H. HALLETT. "The Brighton Waterworks," 8vo, 8 pp. Read at the Brighton Congress of the Sanitary Institute.

Lewes Road Works. By Hollingdean Road.

First well and boring 1830 ? Second well, with galleries, 1853 ?
Engine-room floor 87·85 feet above Ordnance Datum. Level of the bottoms of the headings about 93 feet lower.

Total length of headings 2,150 feet (2,400 according to MR. EASTON). It was rare for 30 feet to be driven without finding a fissure, but the produce of the largest was only from 100 to 150 gallons a minute.

Average daily yield in 1895, 2,000,000 gallons.

Goldstone Bottom Works. Over half a mile northward of West Brighton Railway Station. 1866 ? and later (galleries extended).

Ground-level at the engine-house 147·37 feet above Ordnance Datum. Four shafts. Level of the bottom of the headings about $167\frac{1}{2}$ feet lower. The headings are in north-easterly and north-westerly directions, and about 2,600 feet in length. They vary in size, up to a height of 18 feet and a width of 12 feet.

Average daily yield in 1895, 3,000,000 gallons. Much more at times.

The galleries are in white chalk, with few flints in the flat planes of bedding, but with many oblique layers of thin flint along joint-planes. Some joint-fissures are filled with a soft calcareous sandy deposit, brought down from above by water Some of the chalk seemed fairly soft, but some was found to be hard.

BRIGHTON—BUXTED.   17

The supply comes chiefly from a few large springs a long way apart, yielding from 4,000 to 5,000 gallons a minute, and in connection with joint-planes. There are small additions between these. The contrast between this and the Lewes Road station is remarkable.

In the north-eastern gallery the roof is throughout (1886) of one bed, at the bottom of which was a thin continuous layer of flint, which had been cleared away.

BRIGHTON Industrial School. See FELSCOMBE.

BROADWATER. Rectory.

F. DIXON's "Geology of Sussex," new Ed., 1878, p. 78.

Mould and Gravel - - - - 15 } 22 or 23 feet
Sand, with marine shells of recent species 7 or 8 }

BUXTED. The Box (Mr. E. W. Streeter). 1891.

Made and communicated by MESSRS. A. WILLIAMS & Co.

265 feet above Ordnance Datum.

Shaft 6 feet, then a boring of 6 inches diameter.

Water-level. At the depth of 260 feet, 137 feet down. The boring was then deepened in hope that the water-level would rise. At the last it was 142 feet down, and the yield about 3,000 gallons an hour.

		Thickness.	Depth.
		Feet.	Feet.
[Lower Tunbridge Wells Sand]	Sandstone	11	11
	Hard sandstone	18	29
	Clay and sand	6	35
	Hard white sandstone	2	37
	Sandstone and clay	8	45
	Hard blue clay	8	53
[Wadhurst Clay, 147 feet]	Clay and stone	73	126
	Clay and slate [shale]	8	134
	Clay and stone	49	183
	Stone	2	185
	Clay and stone	7	192
	Hard stone	17	209
	Stone and clay	2	211
	Hard stone	10	221
	Hard blue stone	9	230
	Hard stone and fine white sand	3	233
[Ashdown Sand, 168 feet]	Hard white sandstone	27	260
	Sandstone and white clay	7	267
	Sandstone	30	297
	Hard clay and sand	3	300
	Hard clay	4	304
	Sandstone	20	324
	Sand and clay	10	334
	Sandstone	26	360

An earlier account gives some of the details differently.

A well at the Maypole Inn, north of the village, is 90 feet through clay.

Another, at Pope's Hall Farm, 70½ feet deep, gives an ample supply, the water rising 23 feet.

1178.   B

CATSFIELD. Normanhurst Court. Old Well.
Communicated by MESSRS. TILLEY.
Shaft 145 feet, with adits at the base (? 340 feet long), the rest bored.
? Normal water-level about 110 feet down, lower in summer.
Supply (November, 1886) 2,000 to 3,000 gallons a day, from the bore-hole.

		Thickness.	Depth.
		Feet.	*Feet.*
[Tunbridge Wells Sand]	Sand and loam	12	12
	Blue clay	30	42
[Wadhurst Clay, 136 feet]	Rock with thin layers of blue clay	60	102
	Hard red clay	43	145
	Blue clay	3	148
	Hard white rock	6	154
	Hard white clay	4	158
	Hard white rock	1	159
	Blue clay	8	167
	Red sandstone	2	169
[Ashdown Sand, 69 feet]	Hard white rock	6	175
	Hard blue rock	7	182
	Blue clay	1	183
	White rock	7	190
	Thin layers of coloured [mottled] clay and layers of stone	13	203
	Hard blue clay	14	217

CATSFIELD. For Hastings Waterworks. Just W. of the parish-boundary, a little S. of the north-western corner of Fore Wood.
Communicated by MR. P. H. PALMER, Engineer to Hastings.
45 feet above Ordnance Datum.
Yield (April 1895) about 230,000 gallons in 24 hours, and the 12 months pumping has not affected the springs lower down the valley.

	Thickness.	Depth.
	Feet.	*Feet.*
Alluvial deposit, with much iron-oxide	4½	4½
Grey [Wadhurst] clay	16½	21
Beds of sandrock, with thin layers of clay-shale, dip of about 40° north-eastward	31	52

CHICHESTER. Grayling's Well Farm.
For the Lunatic Asylum, about a mile Northward of the city. 1894.
Boring, made and communicated by MESSRS. DUKE & OCKENDEN.
Lining tubes to 167 feet down. Water-level 50 feet down (October).

	Thickness.	Depth.
	Feet.	*Feet.*
[Drift] Gravel and running sand	20	20
Reading clay beds	80	100
[Upper Chalk] Marl and black flint with small particles of chalk	269	369

CHICHESTER. South Street, Gatehouse's Brewery. 1844.

W. RANGER, "Report to the Local Board of Health, Southampton, on the Various Sources of Water Supply, 1851," p. 48; and SWINDELL and BURNELL, "Rudimentary Treatise on Well-digging," Ed. 4, 1860, pp. 87, 88; and information supplied by Messrs. Gatehouse & Co.

At first water rose so as to yield **26 gallons an hour.** In December, 1845, it yielded 78 gallons; in September, **1846, 90 gallons.** Since then the yield has lessened; in 1885 it was about **45 gallons.** At no time did the water rise to more than 18 feet from **the surface.** The **water** is chalybeate, and smells of **sulphuretted** hydrogen; **its temperature is not** such as to indicate that it rises from the Greensand.

		Thickness.	Depth.
		Feet.	*Feet.*
—	Mould [and made ground]	6	6
[Drift]	Gravel	16½	22½
	Red sand	½	23
[London Clay]	Blue clays	60	83
[Reading Beds]	Coloured (mottled) clays	97	180
	Chalk	600	780
[Chalk, 790 feet]	Crystallized carbonate of lime [Melbourn Rock?]	4	784
	Chalk	125	909
	Chalk marl	61	970
Upper Greensand	Malm Rock containing Iron Stone Nodules	84	1054

CHICHESTER. Waterworks.
Communicated by MR. W. SHELFORD.
Yield 15,000 gallons per hour.

		Thickness.	Depth.
		Feet.	*Feet.*
Loose Soil	—	7	7
[Reading Beds, 18 feet]	Yellow and red clay	9	16
	Black clay, loose, not solid	4	20
	Light-blue clay	1	21
	Marl	4	25
Chalk	—	22	47

CHICHESTER. Westgate.
From notes made during excavation by C. R.

		Thickness.	Depth.
		Feet.	*Feet.*
Drift	Soil	10	10
	Gravel	5	15
	Running gravel		
	Sand	6	21
London Clay	Blue clay	25	46

WATER SUPPLY OF SUSSEX.

CHIDDINGLY. Willowhurst, E. of Stone Cross.
For Major Grant. 1885.
Weald Clay, 112 feet [? more since].

CHILTINGTON. Mr. J. M. Cripps'.
DR. MANTELL, "The Fossils of the South Downs," 4to, London, 1882, p. 84.
Gault. Blue marl, with *Inocerami, Ammonites*, etc., 90 feet.

COOKSBRIDGE. Cottage of Mr. W. Lee.
DR. MANTELL, "The Fossils of the South Downs," 4to, Lond., 1822, pp. 83, 84.
Blue marl, with *Hamites, Ammonites*, etc., 95 } 140 feet.
Marl, with much chlorite [glauconite] sand, 45 }

CRAWLEY. 1898.
Trial-boring, for the Waterworks. About a quarter of a mile south-westward of the Railway Station. 1898.
Communicated by MR. C. O. BLABER.
[Notes in these brackets from specimens.—W W.]
268 feet above Ordnance Datum.

MR. JAMES JOHNSTON adds that water overflowed 12 feet above the ground (small quantity). Pumping 420 galls. per hour, reduced the water-level to 300 feet below surface.

		Thickness.	Depth.
		Feet.	*Feet.*
[Weald Clay]	Clay [brownish]	12	12
	Hard blue clay [light-grey and buff at 15, grey at 46]	34	46
	Soft blue clay [brownish at 52]	6	52
	Rock	11	63
	Blue clay [light-coloured at 84]	33	96
	Undescribed [light-grey clay at 98]	12	108
	Blue clay [brownish-grey at 111]	4	112
	Rock	3½	115½
	Blue clay and rock [greyish clay at 140, darker clay at 146]	32	147½
	Rock [grey clay at 148]	7½	155
	Rock and clay [grey clay at 158, 162 and 173, the last pale]	20	175
	Rock [brownish-grey fissile clay at 185, very pale grey clay at 190]	24½	199½
	Brown rock	1½	201
	Blue and brown rock [grey shaly clay at 204, brownish-grey clay at 206]	8	209
	Rock [grey shaly clay at 210 and 250]	68	277
	Brown rock [light-grey compacted sand at 280]	6	283
	Blue and brown rock	6	289
	Brown rock	32	321
[Tunbridge Wells Series]	Blue rock [light-grey compacted sand at 390]	124	445
	Hard blue rock [very pale grey compacted sand at 500. Very fine grained soft buff earth, compacted, at 550]	139	584
	Sand rock [very light-grey compacted clayey sand at 588]	4 ?	588
	Sand [grey, compacted, ? clayey at 600; grey or buff ditto at 610; light-grey at 630, compacted light-grey, ? clayey, at 637]	60 ?	648

It is clear that the term Rock has been used alike for the firm hard clays and for the fine-grained compacted sands beneath.

In "The Geology of the Weald," Mr. Topley has estimated the total thickness of the Tunbridge Wells Series in this neighbourhood at 380 feet. It seems probable, therefore, that all the beds beneath the Weald belong to this, but unfortunately there is nothing to show the presence of the Cuckfield Clay or of the Grinstead Clay.

For Analysis of the water see p. 107.

CROWBOROUGH, see ROTHERFIELD.

CROWHURST. Just N. of the Furnace Stream (or Asten River), a little W. of S. from the church. Hastings Waterworks.

Communicated by Mr. P. H. Palmer, Engineer.

1. A well.
Water rose to the peat and got away. 210,000 gallons a day.

		Thickness.	Depth.
		Feet.	Feet.
[Alluvium]	Sand and clay	4	4
	Peat and clay	2	6
[Ashdown Beds]	Clayey sand	22	28
	Sandstone	14	42
	Clay	35	77

2. Shaft 62 feet, the rest bored, 10,000 gallons a day run over.

		Thickness.	Depth.
		Feet.	Feet.
Soil		2	2
[Alluvium]	Peat	14	16
[Ashdown Sand]	Yellow sand and traces of clay	17	33
	Clay and sand	7	40
	Fine sand	22	62
	Blue marl (clay)	48	110

3. Fore Wood. Boring. 1898 (? not finished). Yielding upwards of 150,000 gallons a day.

	Thickness.	Depth.
	Feet.	Feet.
Alluvium, with a considerable amount of oxide of iron	4	4
Clayey shale	58	62
Sand-rock	2	64

CUCKFIELD. Workhouse. 1884. 380 feet above Ordnance Datum.
Communicated by Messrs. E. Easton & Co.
Shaft and cylinders 197 feet, the rest bored.

		Thickness.	Depth.
		Feet.	*Feet.*
Upper Tunbridge Wells Sand	Soil.	13	13
	Sandstone	1½	14½
Grinstead Clay	Marl	15½	30
Lower Tunbridge Wells Sand	Sandstone	89	119
Wadhurst Clay	Marl. Very hard thin layers occur at the depth of about 155 to 190 feet. At the depth of 195½ feet a lot of gas bubbled up. At the base the marl is mixed with a little sand	91	200
	Marl, with traces of shells	13	213
	Sand	1	214
	Marl, with traces of shells	5	219
	Very hard sand-rock. A lot of gas met with after passing through this	4	223
	Marl, with traces of shells, and with two inches of very hard sand-rock	27	250
	Marl, with rock at the depth of 282¾ to 284 feet. Very hard rock, 7 inches thick, at 291. 2 inches of rock at about 302. 6 inches of rock at about 309. At the bottom, 10 inches with shells and then 6 inches of rock	63	313
	Marl, mixed with sand. Top, for nearly 6 feet, rock-marl. Then a foot of rock. Rock from 322½ to 326 feet down. 6 inches of rock at about 331.	33	346
Ashdown Sand, 104 feet	Rock	½	346½
	Coarse soft sand	1½	348
	[Undescribed], with hard sand-rock (? ½ foot) at 389	102	450

Another account differs in the details of the Wadhurst Clay, which are thus given :—

	Thickness.	Depth.
	Feet.	*Feet.*
Marl	54	173
Sandstone	1	174
Marl	39	213
Sandstone	1	214
Marl	97	311
Undescribed	35	346

CUCKFIELD PLACE. New Lodge at the entrance to the Park, by the Avenue.
H. W. BRISTOW, in "The Geology of the Weald," p. 93, 1875.
Sand, without water, 25 feet.

DITCHLING. Ditchling Rise, near the Alms Houses.
Bored and communicated by MESSRS. DUKE & OCKENDEN.
Old well, the rest bored - - - - - 70 } 175 ft.
Grey clay and a little sand-rock. No water - 105 }

EASEBOURN. For the Midhurst Rural Sanitary Authority. Just N. of Todham Lock, about 1½ miles E.S.E. of the town.
Communicated by MR. E. EASTON, 1883.
Shaft 11½ feet, the rest bored. Water-level 2 feet down.

	Thickness.	Depth.
	Feet.	*Feet.*
Dark brown clay	1	1
Strong yellow clay	3½	4½
Soft blue clay	1½	6
Blue mottled clay	1½	7½
Marly clay, with sand	4	11½
Charred wood [lignite], with layers of sand	7	18½
Dark sand, with greenish sand	3	21½
Green sand	1	22½
Sandy clay	1½	24
Stiff clay (2 beds)	81½	105½

EASEBOURN (close to Midhurst). About half a mile N.E. of the Workhouse. 1894.
Bored and communicated by MESSRS. LE GRAND & SUTCLIFF.
290 feet above Ordnance Datum.
Water rose 17 feet above the ground.

		Thickness.	Depth.
		Feet.	*Feet*
[Hythe Beds]	Yellow sand and bands of ironstone	18	18
	Sandstone	4	22
	Sand and sandstone in layers	13	35
	Light-coloured clay	3	38
	Clayey sand and pieces of sandstone	26	64
	Clayey sand and bluish stone	76	140
[? Atherfield Clay]	Very fine sand. Large volume of water rose to 13 feet above the surface and overflowed at the rate of 150 gallons a minute at the surface	3	143
	Sandy clay, with nodules	8	151
	Very hard stone	2½	153½
	Sandy clay and stones	9½	163
	Very hard stone	1¾	164¾
	Sandy clay and stones	5¼	170
Atherfield Clay	Stony clay, with fossils	6	176

EAST BLATCHINGTON. Newhaven and Seaford Waterworks, **nearly three quarters of a mile N. of St. Peter's Church.**
Communicated by MESSRS. EASTON & FFOLKES.

159 feet above Ordnance Datum. Shaft 179½ feet, with galleries (N. and S. as well as W. and E.), close to the bottom. Boring of 98 feet a little way in the eastern gallery, to about 90 feet below the bottom of the well. Another about 55 feet along the western gallery, of 145 feet, to about 100 feet below the bottom of the well. Supply from this last.

A bed of flints along the top part of the western gallery **yielded a little water.** A bed near the bottom of the N. and S. gallery gave water in places, but was dry in others.

[Soil, &c.] - - - -	14 or 15 feet
Chalk - - - -	303 or 304 feet

? over 318 feet.

At a visit in 1896, I (W. W.) learnt that **the water-level was** 157 ft. down. There are three shafts and five borings. One of these, of 6 ins. diameter, in the westerly heading, to a depth of 158 below Ordnance Datum, yielded a fair supply, whilst another, of 10 ins. diameter, 3 ft. westward and to a depth of 171 ft. below Ordnance Datum, gave no water. Another, also of 10 ins. diameter, just south of these, in a chamber at the side of the gallery, **to a depth of 217 ft.** below Ordnance Datum, yielded **only a** small supply. One of 8 ins. diameter, to the depth of 115 ft. below Ordnance Datum yielded **no water**; and one of 10 ins. diameter reduced to 4, yielded hardly any. The yield being insufficient, **and the water** having somewhat deteriorated in quality, new works are being **made in Poverty** Bottom, Denton.

For analyses of the water see p. 117.

EASTBOURNE. Star Brewery. 1877.
Sunk and communicated by MR. R. B. PATEN, of St. Albans.
Shaft 50 feet, the rest bored.

		Thickness.	Depth.
		Feet.	Feet.
[Chalk]	Chalk	110	110
	Chalk Marl	53	163
	Marl	39	202
[Upper Greensand]	Rock	1½	203½
	Greensand	1½	205

For analysis of the water see p. 109.

EASTBOURNE. Gas Works (1878?).
From SIR J. PRESTWICH'S MS.

		Thickness.	Depth.
		Feet.	Feet.
[Drift]	Soil and light-coloured clay	2	2
	Yellow sandy clay with angular flints, 2 feet to	4	6
[Gault]	Ash-coloured sandy micaceous clay	8	14
	Gault, with *Baculites, Ammonites,* &c.	173	187

EASTBOURNE. Laundry Company's Works, Latimer-road. 1892.
Made and communicated by Mr. G. Bates, of Lewes. No water.

		Thickness.	Depth.
		Feet.	*Feet.*
Beach		30	30
[Top of Gault ?]	Mixed earth and sand	30	60
[Gault]	Light-blue clay	40	100
	Blue clay	100	200

EASTBOURNE. Lion Brewery.

		Depth.
		Feet.
[Upper Greensand]	Green-grained sandstone (no **water**)	30
	Calcite, with some green-grained sandstone, about ½ inch (with 4,000 gallons an hour)	32½
	Green sandstone (no water)	35

EASTBOURNE. Hygienic **Laundry Company, Upperton Laundry**, Commercial-road. 1893.

A **boring made (in a few** days) and communicated by Messrs. Le Grand & Sutcliff.

Water-level 6 feet down.

		Thickness.	Depth.
		Feet.	*Feet.*
[? Chalk Marl]	Chalk Marl, light-greyish blue	18	18
	Grey chalk	18	36
	Light-blue clay	1	37

EASTBOURNE. Parson's Sawmills. 1885.
Made and communicated by Mr. G. Bates. Good supply of water.

Red earth — 50
Mixed marl and flints — (included)
Chalk and flints — 55
} 105 feet.

EASTBOURNE. Waterworks. Old well, on the marsh northward of the present Engine-house.

Communicated by MR. H. D. SEARLES-WOOD, from a rough section in the office.

Joined by galleries to the newer well, from which pumping is now done.

		Thickness.	Depth.
		Feet.	*Feet.*
Mould		1	1
Alluvium	Yellow clay	13	14
	Peat	1	15
	Blue clay	24	39
Upper Greensand	Sandstone	8	47
	Hard rock	2	49
	Sandstone	10	59

The total depth is given as 100 feet. Probably, therefore, the Gault has been reached. For analysis of the water see page 108.

EASTBOURNE. Waterworks. Newer well with headings, just W. of railway. 1883.

Communicated by MR. J. A. WALLIS (from a drawing at the Works).

Well-top 5·2 feet above Ordnance Datum, and about 15 feet below the level of the ground southward ; less northward, as the ground slopes down to the marsh. This of course adds to the thickness of the Chalk.

		Thickness.	Depth.
		Feet.	*Feet.*
Chalk		25	25
[Upper Greensand]	Green sandstone	3	28
	Hard brown sandstone	2	30
	Green sandstone	24	54
	Hard green sandstone	9	63

EASTBOURNE. Waterworks (see also FOLKINGTON, FRISTON, JEVINGTON, WEST DEAN, and WESTHAM). Trial-boring, by pond northward of Engine-house. 1895-6.

Made and communicated by MESSRS. LE GRAND & SUTCLIFF.
[Notes of specimens.]

		Thickness.	Depth.
		Feet.	*Feet.*
[Alluvium]	Clay	5	5
	Peat	3	8
	Blue Clay	22	30
[Upper Greensand, 35½ ft.]	Greensand and clay [glauconitic]	29½	59½
	Sandstone [glauconitic]	½	60
	Greenish clay and a little sand [whitish and glauconitic]	2	62
	Sandstone [glauconitic]	3½	65½

EASTBOURNE. Waterworks—*continued*.

		Thickness.	Depth.
		Feet.	*Feet.*
[Gault]	Clay and stone [hard, dark sandy clay]	2	67½
	Gault, with **septarium** (6 inches) at base	171½	239
	Gault and fossils [*Inoceramus sulcatus*]	102½	341½
	Gault, green veins and fossils [*Ammonites lautus*]	10	351½
[? Gault and LowerGreensand]	Gault and sand [coarse loamy sand, mixed black and green at 360]	12	363½
	Sand [moderately coarse, with glauconitic grains at 367]	3½	367
	Gault [clay] and sand [coarse sand, and small phosphatic nodules with glauconitic grains at 400]	65	432
Weald Clay	Weald clay [light-grey sandy clay at 432. Dark grey clay at 436. Red mottled clays, specimens down to 510 feet. Whitish silty clay (a 6 inch seam) at 575. Red-mottled clay at 586 and down to bottom]	201	633

EASTERGATE. **Barnham Junction** Railway Station
From a section communicated by the LONDON-BRIGHTON AND SOUTH COAST RAILWAY Co., and from samples down to 233 feet.
About 25 feet above Ordnance Datum.
Sunk 42 feet, the rest bored.

		Thickness.	Depth.
		Feet.	*Feet.*
[Drift]	Yellow sand and stones	12	12
	London clay	44	56
[London Clay, 208 feet]	Sandy loam, with water (1,500 gallons in 24 hours) London clay	56	112
	Rock	1	113
	Blue clay; stiff at 140 feet; sandy blackish with septaria at 150; stiff at 151	38	151
	Rock	1	152
	Blue clay (stiff at 170 feet), sandy with septaria	68½	220½
[Reading Beds, 109½ feet]	Red and mottled clays	108½	329
	Bed of flints	1	330
[Upper Chalk]	Chalk with flints every 3 or 4 feet (a 15 inch bed of flint at 422 feet	105¾	435¾

EASTERGATE. Bognor Waterworks.
Made and communicated by MESSRS. DOCWRA.

Shaft and cylinders 85 feet, the rest bored (24 and then soon 23 inches diameter).

Water-level 25 feet 8 inches down, 14th June 1896.

		Thickness.	Depth.
		Feet.	*Feet.*
Concrete, above the original ground-level		3	3
	Ballast	5	8
	Clay	1½	9½
	Ballast	2	11½
	Marl and Ballast	8	19½
	Clay	2	21½
	Yellow sand	1	22½
	Blue clay	10½	33
	Mottled clay	4	37
[Reading Beds]	Blue clay	3	40
	Mottled clay	14½	54½
	Blue clay	8	62½
	Clay-stones	¾	63¼
	Mottled clay	21½	84¾
	Flints	¼	85
[Upper Chalk]	Chalk and marl	4	89
	Chalk and flints	111	200

EAST GRINSTEAD. Waterworks.
Communicated by MR. E. EASTON, 1883.

Shaft throughout, with galleries at the bottom. Water-level about 29 feet down.

	Thickness.	Depth.
	Feet.	*Feet.*
Clay	10	10
Blue shale	53	63
Red sand-rock	57	120

Another well, for the Gas and Water Co., 1891.
Made and communicated by MR. R. D. BATCHELOR. Shaft throughout.

	Thickness.	Depth.
	Feet.	*Feet.*
Clay, made earth	9	9
Hard dark clay	8½	17½
Septaria (9 inches), and then hard blue clay	2	19½
Rocky sand	7½	27
Rock	2	29
Hard clay	4	33
Rock	3	36
Hard blue clay	6	42
Hard blue shaly clay	15½	57½
Hard rocky sand	7½	65
Hard sand	12	77
Rocky sand	25	102

EAST GRINSTEAD. Brewery.
W. TOPLEY, "Geology of the Weald," p. 86, 1875.

GRINSTEAD CLAY. Blue shale with beds of limestone (probably calc-grit), 70 ft.

The unusual thickness of this clay may be owing to a local flexure, causing the bed to be cut obliquely.

ELSTEAD. On Mr. Albery's land, north of the station.

Made and communicated by MESSRS. DUKE & OCKENDEN.

Water stands 31 feet down.

		Thickness.	Depth.
		Feet.	*Feet.*
[Gault]	Hard dark clay, with 2 inches of rock 50 feet down	95	95
[Folkestone Beds, 139 ft.]	Various sands, green, white, and black, mostly running	105	200
	Sand-rock	2	202
	Various sands, as above	32	234
	Clayey at the base, like pipe-clay.		

FAIRLIGHT. Hastings Waterworks, Ecclesbourne Valley.

Communicated by MR. W. ANDREWS, late Borough Surveyor.

No. 1. Trial Shaft and Boring. On the Northern side of the Fault by the North-eastern end of the reservoir, 1876.

About 250 feet above Ordnance Datum.

Shaft 33 feet, the rest bored (6 inches diameter).

No water found, but some foul air.

		Thickness.	Depth.
		Feet.	*Feet.*
[Wadhurst Clay]	Gravel	10	10
	Dark marl (3 beds)	23	33
	Hard marl	4	37
	Pipe-clay	3	40
[Ashdown Sand]	Hard sand rock (2 beds)	$24\frac{1}{2}$	$64\frac{1}{2}$
	Dark clay	1	$65\frac{1}{2}$
	Hard sand rock	$5\frac{1}{2}$	71
	Dark clay	$2\frac{3}{4}$	$73\frac{3}{4}$
	Hard white sand	$5\frac{1}{4}$	79
	Dark clay	$4\frac{1}{2}$	$83\frac{1}{2}$
	Hard light[-coloured] sand	$3\frac{1}{2}$	87
	Dark clay	5	92
	Hard stone	2	94
	Sand rock	$6\frac{1}{2}$	$100\frac{1}{2}$
	Dark clay	1	$101\frac{1}{2}$

WATER SUPPLY OF SUSSEX.

No. 2. Trial Shaft to 110¾ feet, then bored. About 20 feet lower than the Reservoir, or about 230 above Ordnance Datum (220, W. Topley).

At the depth of 17 feet water flowed in. Also thought at first to flow in at 75 feet 8 inches, and at the rate of about 2,500 gallons a day; but this proved to be soakage.

		Thickness.	Depth.
		Ft. in.	Ft. in.
	Gravel	5 5	5 5
	Dark marl	11 6	16 11
	Gravel	3 7	20 6
	Blue stone	2 8	23 2
	Brown sand rock	0 10	24 0
	Hard stone	0 10	24 10
	Blue marl	3 0	27 10
	Hard stone	0 6	28 4
[? Wadhurst Clay]	Hard blue marl	3 0	31 4
	Shelly stone	0 2	31 6
	Brown marl	0 4	31 10
	Hard blue marl	2 9	34 7
	Hard blue stone	1 6	36 1
	Black sandstone	14 11	51 0
	Brown sandstone	0 6	51 6
	Hard blue stone	3 0	54 6
	Brown sandstone	0 9	55 3
	Dark bind	1 9	57 0
	Very hard rock	1 6	58 6
	Hard blue stone	1 8	60 2
	Sandy bind	4 0	64 2
	Very hard brown rock	0 9	64 11
	Dark sandstone	2 1	67 0
	Very dark sandstone	3 0	70 0
[Ashdown Sand]	White sandstone	5 8	75 8
	Dark sandy bind [No fossils; but ? Endogenites-shale. W. Topley]	13 6	89 2
	Very hard sand rock	2 0	91 2
	White sandy bind	3 0	94 2
	White sandy rock and vegetable deposit	9 0	103 2
	White sandy rock	3 10	107 0
	White sandy rock, with ferns and traces of vegetable deposit	3 9	110 9
	Hard tough ferruginous rock (took eight hours to get through 7 inches)	—	—
	Hard shaly marl or bind, dry at 170 ft.	—	—
	Dry at 177 ft.	—	—
	Drab clay at 189 to 194 ft.	5 0	194 0

FAY GATE. The Beeches (Mr. Frewin's).

Communicated by MESSRS. G. ISLER & Co.

Water-level 28 feet down. Supply abundant.

		Thickness.	Depth.
		Feet.	*Feet.*
Shaft (rest bored)	—	—	32
[Weald Clay]	Rock	1½	33½
	Clay	3½	37
	Clay and rock	12	49
	Clay	3	52
	Clay and rock	9½	61½
	Clay	3¼	64¾
	Rock	6¼	71
	Hard clay and rock	48¾	119¾
	Hard clay	16½	136¼

FAY GATE. Capt. Frazer's (new house).

Bored and communicated by MESSRS. DUKE & OCKENDEN.

		Thickness.	Depth.
		Feet.	*Feet.*
[Weald Clay]	Old Well	—	35
	Hard blue shaly marl and clay	20	55
	Shingle and clay	3	58
	Blue clay	7	65

FILSHAM, see HOLLINGTON.

FISHBOURNE. Opposite the Blacksmith's shop.

Bored and communicated by MESSRS. DUKE & OCKENDEN.

	Thickness.	Depth.
	Feet.	*Feet.*
Dug well	—	20
Clay [Reading Beds]	59	79
Chalk, with good water	20	99

FISHBOURNE. No. 54 Gate-crossing on the railway, a mile west of Chichester Cathedral.

From a tracing communicated by MR. G. L. PURCHASE, City Surveyor.

Shaft 50 ft., the rest bored.

Water-level, December, 1865, about 17½ ft. down.

		Thickness.	Depth.
		Feet.	*Feet.*
	Earth [soil]	1	1
	Sand and gravel	4½	5½
	Light-grey clay	8	13½
	Blue clay with red veins	42½	56
	Blue slate clay	2	58
Reading Beds.	Blue clay with red veins	34	92
106½ ft. ?]	Grey clay	4	96
	Brown clay with blue veins	6	102
	Blue clay	4	106
	Brown clay	6	112
	Chalk, mixed with clay	7	119
	Chalk and flints	4	123
[Chalk, 18 ft.]	Blue clay	3	126
	Chalk and clay, mixed	2	128
	Gravel [? flints]	2	130

It is difficult to say whether there is any London Clay here or not, and therefore it is perhaps safer to class the clays as Reading Beds. Clay may have been carried down into the Chalk in boring, or the blue clay in the Chalk may be one of those marly beds that are not of uncommon occurrence.

FITTLEWORTH, near Pulborough. 1897.

Bored and communicated by MESSRS. DUKE & OCKENDEN.

2 in. tubes to 111 ft. Much ferruginous water at 73 ft.; none below

		Thickness.	Depth.
		Feet.	*Feet.*
	Running sand	53	53
[Hythe Beds]	Sandstone rock	20	73
	Gravel	¼	73¼
	Sandstone rock	1¾	75
	Clay, intermixed with sand [dark sandy clay]	12	87
[Atherfield Clay]	White sand rock	5	92
	Dark sandy clay [with green grains and fossils]	30	122

FOLKINGTON. Trial-boring for the Eastbourne Waterworks, 170 yards east of Broughton Spring. (For adjacent borings, see JEVINGTON, p 56.) Made and communicated by MESSRS. ISLER & Co.
Water-level 25 feet below surface. No supply.

		Thickness.	Depth.
		Feet.	*Feet.*
Dug Pit		—	7
[Rubble?]	Chalk	11	18
[Gault]	Light [coloured] clay	5	23
	Gault	208	231

FOREST ROW. (S.E. of East Grinstead.) Claypits Farm. On the south. Communicated by MR. P. BIRCH (1880). An old well. No water.

		Thickness.	Depth.
		Feet.	*Feet.*
Steining (beds not recorded)		—	15
[Ashdown Sand]	[Fine grained sandstone with vegetable impressions in places.] Tolerably homogeneous, except for bands of soft clay at the depths of about 31, 36, and 67 ft., and about a foot thick	50	65
	Steining (beds not recorded)	5	70
	[Fine-grained sandstone, with occasional vegetable remains]. Apparently interbedded with thin layers of clay at intervals of 3 or 4 inches, and dipping N.E. about 1 in 3 or 4	34	104

FRAMFIELD. Eason's Green, between East Hoathly and Framfield.
W. TOPLEY, "Geology of the Weald," p. 65, 1875.

Wadhurst Clay. Marl (shale), 61½ feet. Good water, probably from the top of the Ashdown Sand.

FRANT. Knowle. S.E. of the village. Colonel H. Grace. 1890.

		Thickness.	Depth.
		Feet.	*Feet.*
[Upper Tunbridge Wells Sand]	Soil	2	2
	White sand	6	8
[Grinstead Clay, 14 feet]	Reddish clay	4	12
	Hard sandstone	6	18
	Clay	1	19
	Reddish clay	3	22
[Lower Tunbridge Wells Sand]	Hard brown rock	1?	23?
	Soft sandstone	6	29

1178.   C

FRANT. Tunbridge Wells Station (South-eastern corner of the yard), London Brighton and South Coast Railway. S. of the town. 1895.

Bored and communicated by MESSRS. LE GRAND & SUTCLIFF.

Fossils determined by MR. E. T. NEWTON.

Water-level 20 feet down. 7,000 gallons an hour easily got. Water flows through a valve at the bottom of the well, fixed to the bore-pipe.

		Thickness.		Depth.	
		Ft.	in.	Ft.	in.
	Old Well (the rest bored). Said to have ended in mottled marl-rock	—	—	89	0
	Dark mottled clay [? Grinstead Clay]	1	0	90	0
	Hard light-blue clay	11	2	101	2
	Blue shaly rock	2	4	103	6
	Blue limy sandstone	3	0	106	6
[Tunbridge Wells Sand]	Dirty white sandstone	5	9	112	3
	Bands of hard buff and irony sandstone	11	9	124	0
	Hard buff sandstone	38	6	162	6
	White sandstone	22	0	184	6
	Hard blue shaly clay	1	6	186	0
	Bands (4 to 9 inches thick) of blue-grey sandstone and blue clay	9	0	195	0
	Hard light-blue marl	1	0	196	0
	Blue clay	0	4	196	4
	Hard blue marl-rock, with ½ inch of granular rock 8 inches down	8	8	205	0
	Hard blue sandy marl-rock	0	6	205	6
	Blue-grey sandstone	0	6	206	0
	Hard light-blue sandy marl-rock	27	0	233	0
	Grey sandstone	3	3	236	3
	Blue sandy marl-rock	11	3	247	6
	Whitish sandstone	5	6	253	0
	Blue sandy marl-rock	7	9	260	9
[? Wadhurst Clay, of great thickness]	Whitish sandstone	2	4	263	1
	Blue sandy marl-rock	3	1	266	2
	Whitish sandstone	1	10	268	0
	Blue marl-rock	13	0	281	0
	Whitish sandstone	1	0	282	0
	Blue marl-rock	4	8	286	8
	Whitish sandstone	9	5	296	1
	Blue marl-rock	1	11	298	0
	Blue marl-rock, with greenish tint	8	0	306	0
	Blue marl-rock	5	3	311	3
	Mottled marl-rock and shale	63	7	374	10
	Dark grey calcareous sandstone	6	6	381	4
	Blue calcareous shaly rock. Fossils from 394 ft. 9 ins. to 398 ft. 3 ins (*Cyrena media*)	26	8	408	0

FRANT. Tunbridge Wells Station—*continued.*

			Thickness.		Depth.	
			Ft.	*in.*	*Ft.*	*in.*
[? Wadhurst Clay, of great thickness]		Blue shaly rock, with layers of sandstone. Fossils from 408 to 410 ft. (*Paludina fluviorum, Cyrena media*) and from 413 to 469¾ ft. (*Cyrena media ?* or *Cyclas*)	10	3	418	3
		Blue shaly rock, with layers of sandstone (no shells)	11	3	429	6
		Light-blue marl-rock	1	3	430	9
		Blue calcareous shaly rock and thin bands of grey sandstone	1	9	432	6
		Light-blue marl-rock	2	0	434	6
		Soft loose blue marl	1	9	436	3
		Blue marl-rock	4	3	440	6
		Blue calcareous shaly rock and bands of grey sandstone	1	10	442	4
		Blue marl-rock	22	4	464	8
		Blue calcareous shaly rock and bands of grey sandstone	3	2	467	10
		Blue marl-rock and frequent bands of ironstone	50	8	518	6
[Ashdown Sand]		Grey sandstone and bands of ironstone	2	0	520	6
		Grey sandstone and thin bands of grey loam	14	4	534	10
		Grey sandstone	9	5	544	3
		Grey sandstone and thin bands of grey loam	4	6	548	9
		Grey sandstone	7	6	556	3
		Grey sandstone and thin bands of grey loam	2	0	558	3
		Grey sandstone	1	3	559	6
		Grey sandstone and thin bands of grey loam	10	0	569	6
		Brown sandy marl-rock	2	6	572	0
		Grey sandstone and bands of grey loam	3	9	575	9
		Brown sandy marl-rock	2	3	578	0
		Grey sandstone and bands of grey loam	1	3	579	3
		Brown sandy marl-rock	0	9	580	0
		Grey sandstone, loam, and sandy marl-rock	7	9	587	9
		Brown sandy marl-rock	1	3	589	0
		Grey sandstone and sandy marl-rock	10	0	599	0
		Hard grey sandstone	3	6	602	6
		Grey marl-rock	2	9	605	3

Should the above classification be right the Wadhurst Clay is of most unexpected thickness, 323½ feet. If, however, the Tunbridge Wells Sand reaches **lower** down than is suggested above (? to 296 feet), then that division **is** of much greater thickness than would have been expected, especially as **the** topmost part is absent. If, again, the Ashdown Sand reaches higher **up** than has been shown, its upper part is exceptionally clayey; but **this is** unlikely, the ironstone often found at **the** base of the Wadhurst Clay **being a** marked bed.

FRANT. Messrs. Wares' Brewery.

Made and communicated by MESSRS. ISLER & Co.

Water-level 85 feet down. Supply, with 3¾ inches pump (barrels 120 feet down), 500 to 600 gallons an hour.

		Thickness.	Depth.
		Feet.	Feet.
[Tunbridge Wells Sand]	Stone and clay	30	30
	Sandstone	103	133
	Blue shale	12	145
	Sandstone	7	152
	Rock	13	165
[Wadhurst Clay]	Brown shale	41	206

FRANT. Rock Cottages, near the south-eastern side of Eridge Park. 1897.

Made and communicated by MESSRS. LE GRAND & SUTCLIFF.

Water-level 60½ feet down.

		Thickness.	Depth.
		Feet.	Feet.
Soil		6½	6½
	Shaly sandstone	½	7
	Blue marl	1	8
	Sandstone and shaly stone	3	11
	Layers of sandstone, blue marl, and clay	4½	15½
	Thin sandstone and coloured clays	7½	23
	Coloured clay	4	27
[? All Tunbridge Wells Sand]	Black shale and stone	5⅝	32⅝
	Sandstone	1⅜	34
	Black shale, stone, and clay, in layers	12½	46½
	Yellow clay and stone	2½	49
	Yellow sandstone	2	51
	Yellow clay and stone, in layers	8	59
	Hard sandstone	8	67
	Yellow clay and stone	2½	69½
	Sandstone	½	70
Tools dropped		5¾	75¾

FRISTON. Eastbourne Waterworks. New Well. 1898.

Communicated by MR. F. STILEMAN.

Shaft, 110 feet; headings, in Upper Chalk, 4,012 feet. Work unfinished. Supply, Dec., 1898, about 6,000,000 gallons per week to the Town, besides what is pumped to waste in the unfinished headings.

(For Analysis of the water, see page 109.)

FUNTINGTON (?). Hambrook House.

Bored and communicated by MESSRS. DUKE & OCKENDEN.

Plenty of water.

[? Reading Beds] {Old well 54; Sand 2} Chalk with flints 79 } 135 feet.

GLYNDE. Lord Hampden's Butter Factory.

Communicated by MR. WELLS.

49½ feet above Ordnance Datum.

Well 6 feet diameter for 50¾ feet; 4½ inch bore to 128¾ feet.

From the sunk well the supply was 1,310 gallons an hour. After completion of the boring the yield was 4,305 gallons an hour. A letter from MR. T. PICKARD (June, 1896) says that the average water-level is 11 feet, and the average quantity pumped 30,000 gallons a day.

		Thickness.	Depth.
		Feet.	Feet.
Lower Chalk	Chalk, much shattered	50¾	50¾
	Chalk [grey marly chalk at 73½ feet; blue sandy marl at 90½ feet]	40	90¾
Unrecorded, probably part Chalk Marl, part Upper Greensand		38	128¾

The sample from 90½ feet corresponds with the lower part of the Chalk Marl. The lowest 2 feet of this well is described as "hard pan," apparently Upper Greensand. For Analysis of the water, see p. 110.

GLYNDE. Mill close to the Railway Station. 1886.

Communicated by MR. T. W. PICKARD.

Average water-level 30 to 40 feet.

Shaft 19 feet, the rest bored. Deepened from 60 feet, later.

Nearly full of water. Overflows in winter.

	Thickness.	Depth.
	Feet.	Feet.
Mould	2	2
Clay, stone-coloured, firm	2½	4½
Black sand, with hazel-nuts and sticks	11½	16
Chalk-rubble	3	19
Chalk-rock	100	119

GROOMBRIDGE. Alongside of Corseley Farm, a quarter of a mile south-west of the Railway Station. 1897.

Made and communicated by MESSRS. LE GRAND & SUTCLIFF.
Water-level 33 feet down.

		Thickness.	Depth.
		Feet.	*Feet.*
Tunbridge Wells Sand]	Marl and sandstone	8	8
	Hard sandstone	2	10
	Hard and soft sandstone	4	14
	Sandy marl	6	20
	Clay and sandstone	$7\frac{1}{2}$	$27\frac{1}{2}$
	Grey sandstone	$6\frac{1}{2}$	34
[? Wadhurst Clay]	Blue clay	3	37
	Sandy blue clay	5	42
	Hard rock	2	44
	Blue clay, sandy	16	60
	Sandstone	2	62

HAILSHAM. Ambergate. [? Amberstone.] Hailsham Water Company.

Shaft, 6 feet diameter.

		Thickness.	Depth.
		Feet.	*Feet.*
Made ground		$2\frac{1}{2}$	$2\frac{1}{2}$
Alluvium	Clayey alluvium	4	$6\frac{1}{2}$
	Alluvium, with tree-trunks, hazel-nuts, etc.	15	$21\frac{1}{2}$
[Tunbridge Wells Sand]	Hard laminated sand-rock	10	$31\frac{1}{2}$
	Soft rock-sand, with some water	5	$36\frac{1}{2}$
	Hard sand-rock	$2\frac{1}{4}$	$38\frac{3}{4}$
	Blue shale	$1\frac{1}{4}$	40
	Very hard sand-rock. Principal source of supply	7	47
	Sandy shale	6	53
	Hard sand-rock	5	58
	Blue shale	4	62
	Hard white rock with some water	$1\frac{1}{2}$	$63\frac{1}{2}$

HAILSHAM. Cottages a mile south of the Railway Station.
Boring made and communicated by Messrs. LE GRAND & SUTCLIFF.
Water level 15 feet 8 inches down.

		Thickness.		Depth.	
		Ft.	in.	Ft.	in.
Dug well	(the rest bored)	—		28	0
[Weald Clay] {	Hard blue clay	4	0	32	0
	Shaly blue clay	41	0	73	0
	Hard blue rock	5	0	78	0
	Blue slaty rock	4	0	82	0
[? Tunbridge	Sand	0	6	82	6
Wells Sand] {	Sandstone	13	6	96	0
	Sand	0	5	96	5
	Rock	3	7	100	0

HAILSHAM. Polegate. Mr. Marsden's. 1876.
Good supply from sand-rock.

	Thickness.	Depth.
	Feet.	Feet.
Old well	65	65
Hard dark clay	125½	190½
Hard sand-rock	6	196½

HARTFIELD. Hartwell. For Mr. J. Mews. 1878.
Sunk and communicated by Messrs. P. DOCWRA & SON.
Shaft 74 feet (?), the rest bored.
Old water-level (? from a different source) 32 feet down; present water-level 52½.

		Thickness.	Depth.
		Feet.	Feet.
Soil		¾	¾
	Sandy marl and sand veins	10	10¾
	Sandy rock	10	20¾
	" " softer	5	25¾
[Ashdown	Hard sand-rock, with veins of sandy marl	5	30¾
Sand]	Rubbly vein and hard sandstone	5	35¾
	Soft sandstone, with veins of white clay	3	38¾
	Sandy marl and hard lumps of rock	4	42¾
	Blue slaty clay, with hard veins	19½	62
[Fairlight	Blue slaty clay, with veins and claystones	10	72
Clays]	Hard blue slaty clay	64	136
	Very hard blue slaty clay	3	139
	Hard blue slaty clay	95	234
	Hard sand-rock	8	242
	Hard light-blue clay	12	254

Another account, also from Messrs. Doowra [? of another well], is as follows:—

Shaft and cylinders 92 feet; water-level, June, 1885, 41½ feet down.

	Thickness.		Depth.	
	Ft.	in.	Ft.	in.
Clay and sand, with beds of stone	59	8	59	8
Brown shaly clay	3	6	63	2
Rock (2 fissures)	3	0	66	2
Blue shaly clay	105	0	171	2
Light-blue clay	9	0	180	2
Hard sand-rock; water 3½ feet down	10	6	190	8
Hard blue shale	4	0	194	8
Light [-coloured] clay	2	6	197	2

Another section (1884).
Shaft 60 feet, the rest bored; water level 20 feet from bottom of well.

	Thickness.	Depth.
	Feet.	Feet.
Made ground	1	1
Congealed sand and loam	10½	11½
Yellow clay, with sand	2½	14
Blue clay	6½	20½
Yellow clay	½	21
Blue clay, with stones	7	28
Clay and sand	6	34
,, more sand	1	35
Blue clay and sand	15	50
Hard stone	1	51
Blue clay, stone, and sand	10	61
Yellow clay and sand	6	67
Blue clay and sand	15	82
Sand	10	92

HARTFIELD. MR. H. B. W. TURNER's.
Made and communicated by Messrs. A. Williams & Co.
Dug pit 10 feet, the rest bored.
Water level 81½ feet down.

		Thickness.	Depth.
		Feet.	Feet.
	Soft clay	2	2
	Sandstone	27	29
	Hard rock	9	38
[Ashdown Sand]	Rock and sandstone	3	41
	Hard rock	17½	58½
	Rock and sandstone	3	61½
	Red sand rock	21½	83
	Red sand rock with marl	8½	91½

HASSOCKS, see KEYMER.

HASTINGS. 41

HASTINGS. Pelham Baths. A boring. 1829.
"The Geology of the Weald," p. 51.
Began nearly at the bottom of the Ashdown Sand.
Beds passed through chiefly clay. Water at 260 feet, rose nearly to the surface.

HASTINGS. St. Leonards Waterworks. (? N. of Caves Road, about a quarter of a mile W. of Church.) 1866.
"The Geology of the Weald," pp. 53-54. 1875.
Shaft 111 feet 2 inches, the rest bored.

			Thickness.		Depth.	
			Ft.	in.	Ft.	in.
Soil, etc.		- - - - -	3	0	3	0
	White sand	- - - -	21	0	24	0
	Brown sand	- - - -	5	0	29	0
	Coarser brown sand	- -	4	0	33	0
	Slaty marl	- - - -	1	0	34	0
	Ferruginous sand-rock	-	4	0	38	0
	Marl, with shells	- -	1	0	39	0
	Strong blue clay	- -	5	0	44	0
	Hard yellow sandstone-rock	-	2	0	46	0
	Grey sandy marl	- -	1	0	47	0
	Variegated marl	- -	1	0	48	0
	Yellow sand-rock	- -	2	0	50	0
	Grey slaty marl	- -	1	0	51	0
	Grey soft stone	- -	3	0	54	0
	Grey marl	- - - -	1	6	55	6
	White marl	- - - -	1	6	57	0
? Ashdown Sand	Grey marl	- - - -	1	0	58	0
	Yellow and blue rock	- -	0	8	58	8
	Soft grey marl	- - -	3	0	61	8
	Marl	- - - -	2	6	64	2
	Hard limestone-rock	- -	4	0	68	2
	Hard blue limestone-rock	-	3	0	71	2
	White clay	- - - -	10	0	81	2
	Grey clay	- - - -	1	0	82	2
	Yellow and blue veined rock	-	1	0	83	2
	Darker blue veined rock	- -	1	0	84	2
	Brown rock	- - - -	0	8	84	10
	Grey hardish rock	- -	3	0	87	10
	Bluish rock	- - - -	6	0	93	10
	Bluish clay	- - - -	2	0	95	10
	Rock	- - - -	0	4	96	2
	Limestone-rock [?calcareous sandstone] -		15	0	111	2
	Strong blue clay	- - -	27	0	138	2
	Clay and sand	- - -	7	0	145	2
	Strong blue clay	- - -	18	0	153	2
	Strong clay mixed with lignite		9	0	162	2
	Hard rock	- - - -	9	0	171	2
? Fairlight Clays, about 173 feet	Strong clay, with a little sand and rock	- - -	12	0	183	2
	Mottled clay	- - -	17	0	200	2
	Sand	- - - -	7	6	207	8
	Rock-sand	- - - -	8	6	216	2
	Mottled clay	- - -	13	0	229	2
	Clay	- - - -	3	0	232	2
	Hard compact clay of various tints, with thin layers of rock - - - - -		51	10	284	0

HASTINGS. Waterworks (see also CATSFIELD, CROWHURST (3), FAIRLIGHT (2), HOLLINGTON (11), and WESTFIELD (8). Less than a quarter of a mile N.N.E. of the remains of the chapel of St. Mary Bulverhithe. Known as the Pepsham or Pepplesham site, being S.S.E. of the farm of that name. Two wells.

Communicated by MR. P. H. PALMER, Engineer.

No. 1. Shaft.

		Thickness.	Depth.
		Feet.	Feet.
Soil		2	2
	Sand and clay	1	3
	Yellow sandstone	13	16
	Clay	$\frac{1}{2}$	$16\frac{1}{2}$
	Hard sandstone	$4\frac{1}{2}$	21
	Grey marl	$1\frac{1}{2}$	$22\frac{1}{2}$
	Hard yellow sandstone	2	$24\frac{1}{2}$
	Hard red sandstone	1	$25\frac{1}{2}$
[Ashdown Sand]	Hard yellow sandstone	1	$26\frac{1}{2}$
	Grey marl	1	$27\frac{1}{2}$
	Yellow sandstone	$2\frac{1}{2}$	30
	White sand and clay, with two inches of clay and sand at top	$4\frac{1}{2}$	$34\frac{1}{2}$
	Grey marl	3	$37\frac{1}{2}$
	Yellow sandstone	$2\frac{1}{2}$	40
	Clay and white sand	3	43

No. 2. Shaft, 60 feet, the rest bored.

		Thickness.		Depth.	
		Ft.	in.	Ft.	in.
	Clay	6	0	6	0
	Sand-rock	33	0	39	0
	Blue marl	1	0	40	0
	Sand-rock	6	0	46	0
	Blue marl	0	10	46	10
	Sand-rock	1	8	48	6
[Ashdown Sand]	Blue marl	2	6	51	0
	Sand-rock	9	0	60	0
	White sand	8	0	68	0
	Clay	1	6	69	6
	Clay and sand	3	6	73	0
	Yellow sand, the bottom part with traces of lignite	7	0	80	0
	Clay	15	0	95	0

HASTINGS. 43

HASTINGS. **Waterworks.** Near the Gasworks, at the **northern part of the town.**
"**The** Geology of the Weald," p. 50. (1875.) **With some additions, from a tracing, and from** MESSRS. TILLEY. **(1883.)**

Shaft 80 feet, the rest bored.

		Thickness.	Depth.
		Ft. in.	Ft. in.
Soil	- - - - - - -	1 0	1 0
	Light-coloured clay	0 9	1 9
	Dark blue clay	2 0	3 9
	Stiff blue clay	4 6	8 3
Wadhurst Clay	Clay and marl, impregnated with iron	1 2	9 5
	Clay and ironstone	3 4	12 9
	Clay, marl, shale and ironstone	5 0	17 9
	Beds of hard stone	6 2	23 11
	Sandstone	4 0	27 11
	Sandstone, with beds of hard stone	8 3	36 2
	Hard **stone**	4 0	40 2
	Sandstone	8 0	48 2
	Hard **stone**	1 0	49 2
	Sandstone, with large open rents	4 0	53 2
	Marl	6 0	59 2
Ashdown Sand, 145¾ feet	Sandstone, with open rents	9 0	68 2
	Marl and stiff blue clay [with *Endogenites erosa*]	22 0	90 2
	Blue clay, with thin beds of hard stone	5 0	95 2
	Hard sandstone	13 3	108 5
	Clay	1 0	109 5
	Sandstone with clay	4 0	113 5
	Stiff blue clay, with thin beds of sandstone	7 7	121 0
	White sandstone	28 6	149 6
	Blue sandy clay	1 0	150 6
	White sandstone	13 0	163 6
	Stiff marly clay [slightly mottled]	5 0	168 6
	Dark sandstone	5 0	173 6
	Stiff blue clay	14 0	187 6
	Clay, with marl [slightly mottled]	7 0	194 6
	Clay, with veins of lignite and vegetable mould	10 0	204 6
Fairlight Beds, 388¼ feet.	Sandy marl	13 0	217 6
	Sandstone	14 0	231 6
	Marl	12 6	244 0
	Sandstone	3 6	247 6
	Dark stiff sandy marl	10 0	257 6
	Dark stiff clay	8 6	266 0
	Hard blue stone	11 6	277 6
	White sand	35 0	312 6
	Marl	3 0	315 6
	Stiff blue clay	14 0	329 6
	Coloured [mottled] clay	16 6	346 0

HASTINGS. Waterworks—*continued*.

		Thickness.	Depth.
		Ft. in.	*Ft. in.*
Fairlight Beds, 388¼ feet.	Light-red sandstone	14 6	360 6
	Dark stiff blue clay	25 0	385 6
	Coloured [mottled] clay, with yellow streaks	10 0	395 6
	Sandy marl, with veins of stone	14 0	409 6
	Dark coloured [mottled] clay	2 0	411 6
	Sandy clay, red in the top part	18 0	429 6
	Pipe-clay	3 0	432 6
	Light-blue clay	5 0	437 6
	Light-coloured clay	6 0	443 6
	Dark red clay	2 0	445 6
	Light-coloured clay	1 0	446 6
	Very dark red clay	3 0	449 6
	Light-coloured clay	6 0	455 6
	Light-coloured sandy clay	18 0	473 6
	White sandstone	1 9	475 3
	Light-coloured clay	1 3	476 6
	Dark red clay	5 0	481 6
	Dark green mottled sandy clay	15 0	496 6
	Dark green clay	8 0	504 6
	Mottled clay	24 0	528 6
	Clay and veins	6 0	534 6
	Clay	17 3	551 9

One account gives the bed next below 257½ feet thus:—

    Stiff clay - - - - - 4½ feet.
    Undescribed - - - - - 8 ,,

and adds 2 or 2½ feet to the depth.

HAYWARDS HEATH. For MR. Bannister. 1883.
Communicated by MR. E. EASTON.
153 feet above Ordnance Datum. Shaft throughout.
Water-level 11 feet down. MR. J. CHURCH says that the water is very ferruginous, and smells.

		Thickness.	Depth.
		Feet.	*Feet.*
Soil	-	1	1
	Sandy loam and clay, with a little soakage	11	12
	Running sandy marl and clay	6	18
	Sand, mixed with blue marl	3	21
	White sand. Strong spring	1½	22½
	Hard blue marl	17½	40
	White sand and marl	1	41
	Hard blue sandy marl	9	50
	Blue marl, nearly as hard as stone	18	68
	Sand	1	69
	Hard blue marl, with occasional small sandy veins	26	95
	White sand. Strong spring	5	100

HAYWARDS HEATH. County Lunatic Asylum.
"The Geology of the Weald," p. 88. 1875. From specimens which had been kept for some time.
MR. TOPLEY notes that in March, 1892, the water was about 10 feet down, and that 40,000 gallons a day were pumped (for analysis see p. 121).

			Thickness.	Depth.
			Feet.	*Feet.*
Soil		-	¾	¾
	White loamy clay	-	5¼	6
	Fine-grained yellow sand	-	5½	11½
	Very fine sand	-	1½	13
	White clay	-	10	23
	Fine white sand	-	4¾	27¾
	Grey and white sand, finely laminated	-	10¼	38
Higher part of Upper Tunbridge Wells Sand	Pipe-clay, with vegetable impressions	-	1	39
	Grey sand, with vegetable impressions	-	4	43
	Sand and conglomerate	-	2	45
	Compact sand	-	15	60
	White sandstone, with carbonaceous specks	-	1	61
	Fine sand, with vegetable impressions	-	9	70
	Iron-pyrites	-	1	71
	Fine loose whitish sand, the bottom 2 feet rather clayey	-	13	84
Cuckfield Clay, 14 feet	Slate-coloured clay	-	2	86
	Sand	-	1	87
	Bluish clay	-	11	98
Lower part of Upper Tunbridge Wells Sand, 69 feet	Fine-grained sand [? rock-bed]	-	16	114
	Blue clay	-	1	115
	Fine-grained sand	-	16	131
	Rather clayey sand	-	36	167
Grinstead Clay, 43 feet	Greenish clay	-	5	172
	Red and green clay	-	2	174
	Red clay	-	2	176
	Clay	-	34	210

HEATHFIELD. For Mr. W. Ash. 1887?
Made and communicated by MESSRS. LE GRAND & SUTCLIFF.
Water-level 35 feet down (September).

		Thickness.		Depth.	
		Ft.	*in.*	*Ft.*	*in.*
[Ashdown Sand]	Clay	20	0	20	0
	Sandstone	6	0	26	0
	Hard marl	2	0	28	0
	Sandstone	4	6	32	6
	Clay	1	6	34	0
	Sandstone	23	6	57	6
	Sandy clay	5	0	62	6
	Sandstone	6	10	69	4
	Sandy clay	12	2	81	6
	Sandstone	0	6	82	0

HEATHFIELD Station and Hotel, see WALDRON.

HELLINGLY. Park Farm. For the Sussex County Council.
Made and communicated by MR. A. E. NUNN, and from MR. C. O. BLABER
110 feet above Ordnance Datum.
Shaft 76 feet, the rest a boring of 10 inches diameter.
Water rises to 72 feet from the surface.

	Thickness.	Depth.
	Feet.	Feet.
Sandy soil	4	4
Red clay	8	12
White hard clay	8	20
Blue marl	7	27
White hard sand	6	33
Black hard sand	5	38
White hard sand	6½	44½
Blue marl	4	48½
Grey sand	4½	53
Brown sand	7½	60½
Blue marl	22½	83
Grey sand	37	120
Blue clay	3	123
Hard rock	7	130
Purple brown clay	36	166
Sand rock	12	178

HENFIELD. General Gordon's. 1895.
Boring, made and communicated by MESSRS. DUKE & OCKENDEN.
Water rises to 56 feet from the surface. (For analysis see page 113.)

		Thickness.	Depth.
		Feet.	Feet.
[Folkestone Beds]	Running sand and water	25	25
	Clay (thin rock at 149)	—	to 163

HENFIELD. "Gardner's Arms." 1899.
Bored and samples communicated by MESSRS. DUKE & OCKENDEN.
Water, from the Folkestone Beds, rises to 47 feet from the surface.

		Thickness.	Depth.
		Feet.	Feet.
[Folkestone Beds]	Yellow sand [ferruginous sand and iron sandstone at 1, 11, and 17]	24	24
	Running sand [glauconitic at 30]	33	57
[Sandgate Beds]	Blue clay intermixed with sand [dark-green clayey sand at 106 ; dark sandy clay at 111 ; sandy clay and pyrites at 125, 127 ; black clay at 137]	83	140
	Grey fuller's earth	4	144
	Dark-green loamy sand	—	—

At the time of going to press this boring was still unfinished, and the Hythe Beds do not appear to have been reached. Another boring, at the Bull Inn, made in 1893, reached 200 feet, in "blue clay," apparently without touching Hythe Beds.

HOLLINGTON. For Hastings Waterworks. Filsham or Bopeep.
Boring No. 1. East of the Marsh (? finished 1881).
Communicated by Mr. E. Easton.

		Thickness.	Depth.
		Feet	*Feet.*
Soil		4	4
[Ashdown Sand]	White sand (3 beds)	28	32
	Brown sandstone, top 5 feet soft	20	52
	Dark sand and clay	10	62
	Brown sandstone and clay	16	78
	Brown sandstone	10	88
	Light-blue pipe-clay	11	99
	Dark sandy clay	6	105
	Dark clay, with lignite	2	107
	Dark blue clay	3	110
	Blue clay, with thin beds of sandstone and vegetable remains	11	121
[? Ashdown or Fairlight Beds]	Blue clay, with thin layers of lignite	10	131
	Blue clay	9	140
	Light-blue clay	14	154
	Sandy clay	6	160
	Sandy clay and pebbles	8	168
	Blue clay and sand	7	175
	Blue clay and lignite	11	186
	Sandy clay and lignite	14	200
	Sandy clay	12	212
	Light [-coloured] sandy clay	11	223
	Dark sandy clay (2 beds)	27	250

[This points to a slight extension of the outcrop of the Ashdown Sand, as shown on the Geological Survey Map, and perhaps also to the occurrence of Fairlight Beds nearer the surface than would have been expected.]

Filsham (No. 1 Well), also from Mr. E. Easton, is as follows :—

Shaft 64 feet, the rest bored.

		Thickness.	Depth.
		Feet.	*Feet.*
Soil		4	4
[Ashdown Sand]	White sand	30	34
	Brown sandstone	19	53
	Dark sandy clay	10	63
	Brown sandstone	4	67
	Brown sand with beds of clay	13	80
	Pipe-clay	9½	89½

Filsham (No. 2 Well), at a slightly higher level.
Shaft, with a heading about 60 feet down.

		Thickness.	Depth.
		Feet.	Feet.
Soil		6	6
[Ashdown Sand]	White sand	20¾	26¾
	Light-brown sand	9½	36
	Dark-brown sand	10	46
	Light-brown sand	8	54
	Light [-coloured] sandy clay	10½	64½
	Very dark bind	1¼	65¾
	Hard sand-rock	5	70¾
	Pipe-clay	1	71¾
	White sandy clay	10¼	82

In the upper part the beds dip 1 in 6 to the west. Lower down (at the depth of about 40 feet ?) 1 in 12.

Of wells Nos. 3 and 4 there is no record.

Filsham. Boring No. 5. In the marsh, N.W of the Pumping Station.

Communicated by Mr. P. H. PALMER, Engineer (and from specimens. The colours much alike throughout, mostly a sort of brownish grey or buff).

		Thickness.	Depth.
		Feet.	Feet.
[Alluvium]	Soil	2	2
	Clayey soil	4	6
	Peat	12	18
	Blue clay	5	23
	Peat	2	25
	Blue muddy clay	5	30
[Looks as if the Ashdown Sand was absent, and Fairlight Clays near the surface]	Muddy clay, with fine traces of sand [pale brownish-grey clay, with traces of twigs and slight streaks of Vivianite]	29½	59½
	Yellow clay, with small stones [? a fine gravel]	½	60
	Clay [buff, sandy]	3	63
	Yellow clay, with small stones [? a very fine gravel or broken up stone]	1	64
	Dense blue clay [buff fine clayey sand or sandy clay. *Endogenites?*]	24	88
	Clay [buff sandy] with slight traces of sand	8¾	96¾
	Clay [buff sandy], with traces of sand and small stones	4½	101
	Clay [pale-grey] with traces of (buff) sand	5½	106½
	Clay [sandy] and lignite [streaks]	½	107
	Clay [pale grey] and sand	2	109

HOLLINGTON.

Filsham. Boring No. 5—*continued*.

		Thickness.	Depth.
		Feet.	*Feet.*
	Dense clay [buff], with traces of lignite	10	119
	Clay [buff sandy] and sand	11	130
	Clay [buff clayey sand or sandy clay] and lignite	1	131
	Clay [sandy and clayey sand]	32	163
	Soft white sand [pale grey, fine]	1½	164½
	Clayey sand [blackened] with lignite	½	165
	Sandy clay [grey]	9	174
	Clay, with white [buff] sand	2	176
	Clay [pale grey] and sand [paler]	2	178
	Clay [pale grey, streaked with paler sand]	18	196
	Clay [light-grey, sandy]	22	218
	Clay [pale grey or buff, sandy]	13½	231½
[Looks as if the Ashdown Sand was absent, and Fairlight Clays near the surface.]	Pipe-clay [pale grey or buff, sandy]	9½	241
	Black clay and sand [clayey sand, coloured by lignite]	1	242
	Black clay and sand, with lignite [pale grey]	2	244
	Sand [fine, sharp, light-coloured, with grains of lignite (? from above)]	7	251
	Clay [pale grey sandy clay or clayey sand]	2	253
	[Brownish-grey clay at 267]	14	267
	[Brownish-grey and light-coloured clay and sandy clay at 270]	3	270
	[Pale grey clay at 272]	2	272
	[Brownish-grey sandy clay at 306. Pale-grey clay at 308]	36	308
	[Stiff grey and crimson-mottled clay at 311]	3	311
	[Grey clay at 329. Grey and buff clay at 334. Brownish-grey clay at 341. Buff and pale-grey sandy clay at 350]	41	352

HOLLINGTON. Hastings Waterworks. **Wells near Old Roar or** Buckshole Reservoir. [Between this and Harmer's **Reservoir.** No. 3, about half way. Nos. 2, 1, and 4, successively nearer the latter.]

No. 1. 54·2 feet above Ordnance **Datum.**
Shaft of 69 feet. Water-level 59 **feet down.**
The water, if left to itself, before the heading was made, overflowed, according to MR. W. ANDREWS.
Connected with No. 2 by a heading. No details.

No. 2. 66¼ feet above Ordnance Datum.
Shaft **71 feet,** boring 95. Water-level **33½** feet down.
Two springs, from W. and S.W. (W. Topley). No details.
Yield of 1 and 2, 87,000 gallons in 24 hours, according to MR. W. ANDREWS (1875).

No. 3. 82·9 feet above Ordnance Datum.
Shaft 88½ feet, the rest bored.
Water found 47½ feet down (35,000 gallons a day. Enters from N.W. and S.W W Topley).

——	——	Thickness.	Depth.
		Feet.	*Feet.*
	Blue clay and gravel	20	20
	Blue clay	10	30
[? Ashdown Sand]	Sandstone	4	34
	Blue clay	10	44
	Hard grey rock	3½	47½
	Sandstone	1¾	49¼
	Blue bind	5¼	54½
	Mingled blue and red bind	2	56½
	Grey bind	2	58½
	Sandstone	6	64½
	Black bind	5½	70
	Blue bind	53½	123½
	Sandy soil	11½	135
	Grey bind	5	140
	Sandstone	7	147
	Blue bind	10	157
	Sandstone	16	173
	Blue bind	16¼	189¼

No. 4. About 51·9 feet above Ordnance Datum. 1875.
? Shaft 112 feet, with a heading of 120 feet, ? the rest bored. Yield about 80,000 gallons in 24 hours, according to Mr W. Andrews (1875). [Words in these brackets by W. Topley, as also particulars below 102 feet.]

——	Thickness.	Depth.
	Feet.	*Feet.*
Bog	17	17
Grey bind [sandy clay]	15	32
Sandy soil [laminated sandstone]	10	42
Dark bind [sandy clay]	9	51
Sandy bind [laminated]. Water at the depth of 62¼ feet [28,000 gallons a day]	13¼	64¼
Sandstone [white]	4½	68¾
Hard grey rock [coarse]	5¾	74½
Black bind [sandy]	2½	77
Sandstone [white, clayey]	3¼	80
Dark bind and lignite	2¾	83
Hard sandy bind	3	86
Blue bind	2	88
Mingled blue and red bind	2	90
Blue bind	9	99
Mingled blue and red bind [pisolitic iron bed]	3	102
Hard white sandstone. Water at the depth of about 110 feet ; 60,000 gallons a day	?13½	115½
Mottled clay [? soon passing down into] blue bind	?49½	165
Fine sand-rock	7	172
Soft blue bind	4	176

No. 5. About 30 feet beyond the heading of No. 4. From a letter by Mr. W. Andrews, August 1876. Large spring found in this shaft (August 1876) at the depth of about 90 feet, when air and water rushed up with noise, giving the workmen barely time to get out of the well. Water rose 60 feet

HOLLINGTON.

in 1½ hours, and seems to be almost stationary at 64 feet. Beds different from those in the other shafts; nearly the whole a rotten blue marl, dip opposite to that in No. 4 well and heading (some at angle of 45°). At 72 feet part of a fossil fish.

Of No. 6 there is no record.

No. 7. Old Roar Valley. 1880.

[Words and figures in brackets from a tracing from MR. W. ANDREWS, Surveyor.]

Measurements from the level of the pump. [Shaft, 100 feet, the rest bored.] Measurements originally taken from the top of the well-frame, which is 2 feet above the ground level. This has been altered to the ground level here.

	Thickness.	Depth.
	Ft. in.	Ft. in.
Loam and sandy ground [17]	10 0	10 0
Loose black bind [16]	23 0	33 0
Hard black shale	5 0	38 0
Green bind	12 0	50 0
Black bind	10 0	60 0
Black shale	6 0	66 0
[Beds partly on end (fault. W. Topley). 3.]		
Hard blue stone [high dip, seems to cut across the other beds]	4 0	70 0
Blue bind (? 2 beds)	9 0	79 0
Black shale	18 3	97 3
Blue-stone	1 3	98 6
Sandstone [? 2 beds, 4 feet 2 inches]	2 2	100 8
Blue stone and clay	7 4	108 0
Dark grey bind	8 4	116 4
Bluish sandstone	1 8	118 0
Fine light (-coloured) sandstone	1 0	119 0
Bluish sandstone	2 0	121 0
Dark sandy bind [thin beds, black and white]	3 0	124 0
Black bind	3 0	127 0
Black clay bind and thin bedded sandstone	23 0	150 0
White sandy clay	2 0	152 0
Dark sandy bind	35 0	187 0
Dark brown bind	1 6	188 6
Dark brown sandstone	2 0	190 6
Light-blue pipe-clay	2 6	193 0
Light (-coloured) soft sandstone	9 0	202 0
White sand [the drill went part through this]	9 6	211 6
Light (-coloured) sandy clay and lignite	8 6	220 0
Dark clay and sand, hard	10 0	230 0
Brownish bind	17 0	249 0
Brown sandstone and lignite	0 6	249 6
Light brown sandy clay and vegetable remains (lignite) and then layers of sandstone	16 6	265 0
Light [-coloured] sandy, brown and yellow bind with thin beds of sandstone and vegetable remains (lignite)	15 0	280 0
Sandstone	1 0	281 0
Light [-coloured] and brown clay	1 0	282 0
Blue pipe-clay	11 6	293 6
Light [-coloured] sandstone	6 6	300 0
Very fine soft green sandstone	8 0	308 0
Coarse light [-coloured] sandstone	2 0	312 0
Mottled clay bind, red, blue and yellow	10 0	324 0
Very fine light [-coloured] sandstone	7 0	339 0
Dark brown bind	0 6	339 6
Mottled clay bind, red, blue and yellow	10 6	340 0

The beds have a slight dip northward. A good deal of water was met with at the depth of 98 to 102 feet 8 inches, supposed to be a spring connected with Nos. 6 and 4 wells.

HOLLINGTON. Hastings Waterworks, originally Mr. Burton's. Two wells connected by a heading. At the back road over a sixth of a mile southward of St. John's Church. 1874.

From a note by MR. TOPLEY.

(?) Shaft 100 feet, the rest bored.

Yield 70,000 to 75,000 gallons a day. According to a letter from MR. BURTON (Jan. 1875), this supply rises out of the borehole into the well (for analysis of the water see page 113).

Wadhurst Clay - - - - - - 24 } 141 feet.
Ashdown Sand. Sandstone and sand - - 117 }

HOLLINGTON. Silver Hill. For the Hastings Rural Sanitary Authority. On the western side of the high road over a sixth of a mile northward of St. Matthew's Church. 1885.

Communicated by MESSRS. JEFFERY & SKILLER.

? About 225 feet above Ordnance Datum.

Shaft 215 feet, the rest bored

A later letter (1889) adds that water was found at 215 feet, the calculated yield being about 80,000 gallons in 24 hours. Supply 40,000 to 50,000 gallons a day, with no falling off in the yield.

		Thickness.	Depth.
		Feet.	Feet.
[Wadhurst Clay]	Yellow clay marl - - -	16	16
	Blue marl - - - -	16½	32½
	Blue stone - - - -	5	37½
	Blue marl - - - -	44½	82
	Blue stone - - - -	4	86
[? Ashdown Sand]	Yellow sandstone - - -	5	91
	Brown sandstone - - -	4	95
	Yellow sandstone - - -	5	100
	Blue marl, changing to following - - - about	95	195
	Yellowish sand-rock, in which the headings were driven about	30	225

There are now (1894) two wells 50 feet apart, connected by a heading. Another heading runs 100 feet westward from the newer well, at the bottom of which is a boring, in which the jumpers were lost in a fissure.

HORSHAM.

Horsham. Stammerham. Christ's Hospital.

Made and communicated by Messrs. **Docwra.**

Shaft and cylinders **140½ feet**, the rest bored.

Water-level 57 feet 10 inches down, October 20, 1896.

	Thickness.	Depth.
	Ft. in.	Ft. in.
Sandstone rock (with girder, 7 inches, above)	65 0	65 0
Blue shale	30 0	95 0
Red marl	2 0	97 0
Blue shale	74 0	171 0
Red marl	1 0	172 0
Blue shale	40 0	212 0
Hard beds of ironstone rock and blue shale	81 7	293 7
Hard sand-rock	16 8	310 3
Hard clay	2 6	312 9
Hard sand-rock	38 3	351 0
Blue shale, with couts rock from 353 to 353¼	3 0	354 0
Hard sand-rock, with couts rock from 400 feet 6 inches to 401 feet 2 inches	63 0	417 0

Horsham. Waterworks, a little W. of Railway Station.

Communicated by Mr. P. **Chasemore.** 1890.

Shaft 74 feet, the rest bored (? more than one bore-hole.)

		Thickness.	Depth.
		Feet.	Feet.
	Clay - - - about	4	4
[? Upper Tunbridge Wells Sand]	Srave or shrave, about	10	14
	Brown marl	46	60
	Rock of sand	8	68
	Blue marl	6	74
	Land-rock [? sand-rock] about	8	82
[? Grinstead Clay]	Blue marl	20	102
	Rock	3	105
	Blue marl	18	123
[? Lower Tunbridge Wells Sand]	Very hard rock, under which water was found, "between two rocks about nine inches apart"	¾	123¾

Horsham. London and Brighton Railway Station.

Well, 1881. Boring, 1895. Shaft 91½ feet, the rest bored.

The well from a small drawing (in Mr. Topley's collection), from a letter from Mr. R. J. Billington, and from a letter from Mr. P. Neate. The boring made and communicated by Messrs. Le Grand & Sutcliff.

The following particulars from the Railway Co. Two headings, 10 feet high, 6 feet wide and 40 long, bottom about 87 feet down. The well **at first** yielded about 25,000 gallons in 24 hours, but since the Water Company has deepened its well the effect has been a shorter **supply** in this.

Water-level 37½ feet down.

	Thickness.	Depth.
	Feet.	*Feet.*
Soft clay	5	5
Hard sandstone. Impure limestone in another account	14	19
Soft blue marl	4	23
Hard blue limestone (sandstone in note). Finely bedded clayey sand in another account	16	39
Soft blue marl	1	40
Blue rock. Finely bedded clayey sand in another account	2	42
Soft blue marl	3	45
Hard blue limestone (sandstone in note). Finely bedded clayey sand in another account	30	75
Hard sandstone. Described as flaggy and micaceous in another account	1	76
Hard blue limestone. Finely bedded clayey sand in another account	13	89
Unaccounted for, Messrs. Legrand & Sutcliff giving the depth of the shaft as 91½	2½	91½
Grey sandstone and blue marl rock	9½	101
Thin bands of sandstone and blue marl rock	11½	112½
Bands of sandstone and blue marl rock	22½	135
Grey sandstone	8	143
Grey sandstone and blue marl rock	4	147
Grey sandstone	4	151
Bands of grey sandstone and blue marl rock	24½	175½
Bands of grey sandstone and thin bands of blue marl rock	5	180½
Grey sandstone	8	188½
Bands of grey sandstone and blue marl rock	9½	198
Grey sandstone	1	199
Blue marl rock	5	204
Bands of grey sandstone and blue marl rock	1	205
Mottled marl rock	7½	212½
Sandy marl rock	5½	218
Mottled marl rock	8	226
Grey sandstone	35	261
Hard blue marl rock	6	267
Grey sandstone	9½	276½
Bands of grey sandstone and blue marl rock	12½	289
Mottled marl rock	8	297
Hard grey sandstone	8	305
Mottled marl rock	17	322
Grey sandy marl rock	9½	331½

The beds may be all part of the Tunbridge Wells Sand, the various and more or less local divisions of which, however, one cannot make out. The lowest 26½ feet may, however, belong to the Wadhurst Clay.

For Analysis of the water see p. 114.

HORSTED KEYNES. Railway Station. 1896.
Boring made and communicated by Messrs. Le Grand & Sutcliff.
Water-level 6 feet 9 inches down.

			Thickness.	Depth.
			Feet.	*Feet.*
Dug well (the rest bored)			—	45
[Wadhurst Clay?]	{	Blue marl and bands of blue shale	36	81
		Blue marl and bands of shale and ironstone-nodules	4	85
		Blue marl and bands of shelly limestone	26	111
		Blue marl-rock, bands of shale, ironstone-nodules, and bands of blue-grey sandstone	3	114
[All Ashdown Sand]	{	Bluish-grey sandstone and bands of blue marl-rock	$7\frac{1}{2}$	$121\frac{1}{2}$
		Grey sandstone and thin bands of blue marl-rock	$45\frac{1}{2}$	167
		Grey marl-rock	1	168
		Grey sandstone and bands of grey marl-rock	25	193
		Grey marl-rock	$6\frac{1}{2}$	$199\frac{1}{2}$
		Grey marl-rock and thin bands of grey sandstone	$10\frac{1}{2}$	210
		Grey sandstone	$25\frac{1}{2}$	$235\frac{1}{2}$
		Grey sandstone and bands of blue marl-rock	17	$252\frac{1}{2}$
		Blue marl-rock	2	$254\frac{1}{2}$
		Grey marl-rock	$1\frac{1}{2}$	256
		Grey sandstone	$2\frac{1}{2}$	$258\frac{1}{2}$
		Grey marl-rock	2	$260\frac{1}{2}$
		Grey sandstone	$2\frac{1}{2}$	263
		Brown marl-rock and lignite	$2\frac{1}{2}$	$265\frac{1}{2}$
		Grey sandstone and bands of blue marl-rock	$18\frac{1}{2}$	284
		Blue marl-rock	$7\frac{1}{2}$	$291\frac{1}{2}$
		Brown marl-rock	$11\frac{1}{2}$	303
		Brown marl-rock and bands of grey sandstone	$6\frac{1}{2}$	$309\frac{1}{2}$
		Brownish mottled marl-rock and bands of grey sandstone	9	$318\frac{1}{2}$
		Grey sandstone	3	$321\frac{1}{2}$
		Brownish mottled marl-rock and bands of grey sandstone	$5\frac{1}{2}$	327
		Grey sandstone and bands of blue marl-rock	16	343
		Brown marl-rock and bands of grey sandstone	$5\frac{1}{2}$	$348\frac{1}{2}$
		Grey sandstone	11	$359\frac{1}{2}$
		Blue marl-rock	$1\frac{1}{2}$	361
		Grey sandstone and bands of blue marl-rock	$12\frac{1}{2}$	$373\frac{1}{2}$
		Brown marl-rock	$4\frac{1}{2}$	378
		Brown marl-rock and lignite	$7\frac{1}{2}$	$385\frac{1}{2}$
		Brown marl-rock	$1\frac{1}{2}$	387
		Grey sandstone	$20\frac{1}{2}$	$407\frac{1}{2}$

HUNSTON. Hoe Farm.

Sunk and communicated, from memory, by MR. OCKENDEN, SENR.

		Thickness.	Depth.
		Feet.	*Feet.*
[Bracklesham Beds]	Sand	42	42
[London Clay and Reading Beds]	Clay, lower part red	233	275
Chalk		1½	276½

JEVINGTON. Trial borings for the Eastbourne Waterworks. 1896.

(For another adjoining see FOLKINGTON).

Made and partly communicated by MESSRS. ISLER, and partly by MR. F. STILEMAN, and from specimens and observations.

1. Shaft **7 feet, the rest bored.** About 100 yards, a little E. of N. of the outbreak of Broughton Spring, which is over half a mile E. S. E. of Folkington Church. Surface 99·5 feet above Ordnance Datum.

		Thickness.	Depth.
		Feet.	*Feet.*
Chalk [wash of chalk down slope]		5	5
[Gault, 345 feet]	Blue marl (specimens dark grey sandy clay, 60 and 68 feet)	275	280
	Greensand and gault, mixed from clay falling in (specimen, green sandy clay, with some grey clay)	54	334
	Dead greensand (specimen, like the above, with very little clay	16	350
[Weald Clay]	Brown clay (specimen, grey with brownish patches)	4	354
	Blue clay (specimen, dark grey)	7	361
	Light [-coloured] clay (specimen, mottled grey and brownish)	3	364
	Blue clay (specimen, grey with a little brownish	30	394
	Red and grey mottled clay	11	405
	Dark greenish-grey clay	8	413

The beds between 280 and 350 feet perhaps represent the Lower Greensand.

2. About 50 yards a little N. of W. from the head of Broughton Spring. Shaft 49 feet, the rest bored. Surface 108·5 feet above Ordnance Datum.

		Thickness.	Depth.
		Feet.	*Feet.*
Gault	Top soil [and rainwash?]	14½	14½
	Upper Greensand	4	18½
	Specimen, dark grey clay, said to be all alike	184½	206
	Sandstone	2	208
	Gault	30	238

JEVINGTON. Cottage on the eastern side of the road, close to the southern end of Wannock Coppice. 1880.

Information from MR. MILLER, foreman to Mr. Diplock, the owner.

Shaft 102 feet, the rest bored.

On the completion of the boring water rose 36 feet, at the rate of 2 feet an hour, and then continued rising (? at less rate) to within 15 feet of the surface.

		Thickness.	Depth.
		Feet.	*Feet.*
[Gault]	Clay, with a foot of rock at the base	12 ?	12
	Hard clay, with shells. Another rock about 56 ft. down	90	102
	Clay, to sand	56 ?	158

Further information from MR. F. STILEMAN gives the following data:—
Level of coping 73·9 feet above Ordnance Datum.
Bottom of well 21·6 feet below „ „
Water-level 53·8 feet above „ „
This seems to show that the well may have been silted up somewhat.

KEYMER. Hassocks.

New house on the Brighton Road, about 27 chains south of the road to the Station.

From information and samples obtained during the work (1890).

		Thickness.	Depth.
		Feet.	*Feet.*
[Gault]	Weathered clay	12	12
	Black clay	18	30
	Coarse greensand, mixed with clay (water)	9	39

KEYMER. Hassocks Gate.

At Mr. Stevens', close to Hassocks Goods Station, on the west side of the line.

Bored and communicated by MESSRS. DUKE & OCKENDEN.

[Lower Greensand] Black clay, rock, and sand, to red sand with water, 40 ft.

KEYMER. Leylands Park, just W. of Keymer Junction Station. 1890.
Made and communicated by MESSRS. LE GRAND & SUTCLIFF.
114 feet above Ordnance Datum. Water-level 12½ feet down.

		Thickness.		Depth.	
		Ft.	in.	Ft.	in.
Soil		1	0	1	0
[Weald Clay]	Weald clay	134	6	135	6
	Hard Weald clay and sand conglomerate	5	6	141	0
	Weald clay	2	0	143	0
	Weald clay and sand	5	0	148	0
	Weald clay	3	0	151	0
	Weald clay, slightly mottled	6	0	157	0
	Weald clay	22	6	179	6
	Weald clay, slightly mottled, with a rock-band 3 feet down	5	0	184	6
	Rock	9	6	194	0
	Clay	19	6	213	6
	Hard clay and sand	8	0	221	6
	Hard clay	31	6	253	0
	Hard clay and sand	50	0	303	0
	Hard clay	1	3	304	3
	Very hard clay, with green specks	2	3	306	6
	Very hard clay	25	3	331	9
	Very hard clay, mottled	2	10	334	7
	Very hard clay	21	8	356	3
	Red mottled clay	3	1	359	4
	Dark hard clay	8	4	367	8
	Hard clay and brownish sand	8	9	376	5

KINGSTON. Newmarket. 1893.

Made and communicated by MR. G. BATES. Good supply of water.

Chalk - - 50 ⎫
Chalk and flints 50 ⎬ 100 feet.

KIRDFORD. At Fittleworth Scrub House.

P. J. MARTIN. "A Geological Memoir on a Part of Western Sussex."
pp. 42, 43. 4to, London. 1828.

		Thickness.	Depth.
		Feet.	Feet.
[Weald Clay]	Reddish clay	12	12
	Marble	1½	13½
	Blue clay and shale, with much selenite; also *Cypris* and fish-scales, hardened and often passing into fuller's earth	44	57½

LAMBERHURST. Brewery.
Made and communicated by Messrs. Docwra.
Water-level 22 feet down.

		Thickness.	Depth.
		Feet.	Feet.
	Shaft (the rest bored), undescribed	—	35
	Sandstone rock	10	45
[Tunbridge Wells Sand]	Gault [clay]	2½	47½
	Sandstone rock	7½	55
	Gault [clay]	3½	58½
	Sandstone rock	10	68½
	Gault [clay]	3	71½
	Black slaty rock	7	78½
[Wadhurst Clay]	Gault [clay]	5	83½
	Hard slaty rock	6	89½
	Gault [clay]	2½	92
	Hard slaty rock	2½	94½*

* Given as 95½

LANCING. In the Level. ? About 300 yards east of Lower Lancing.
F. Dixon's "Geology of Sussex," Edition 2, 1878, p. 77.

	Feet.
Rolled flints and sand	8 or 10
Marl	10 or 12
Upper Chalk, with flints, and with an excellent supply of water	5 or 6

LANCING. The Terrace. 1891.

Made and communicated by Messrs. Le Grand & Sutcliff.
Water-level 16 feet down.

		Thickness.	Depth.
		Feet.	Feet.
	Made ground	6	6
	Gravel	13	19
	Mottled clay	12	31
	Dark clay	4	35
[? Reading Beds]	Black shaly clay	9	44
	Mottled clay	4	48
	Dark clay, with boulders [? flints]	6	54
	Chalk and flints	41	95

LAUGHTON. Laughton Place.

Dr. Mantell. "The Fossils of the South Downs," 4to, London, 1822, p. 82.
Gault. Blue marl, with many fossils 60 feet.

LEWES. The Baths.

Waterworks Investment Review, January, 1898.

At the depth of 40 feet water rose to within 12 feet of the surface. After sinking the pipe a few feet lower it rose to within 8 feet of the surface.

		Depth.
		Feet.
[Alluvium]	To blue-grey clay	10
	Alluvial deposit found at	20
	Greyish-green clay	—
	A sort of brown sand at	29
	A green sand at	31
	Blue gault at	34
	A foot layer of petrified wood at	36
	Flints, 2 feet	—
	To Chalk	39
	In Chalk to	59

LEWES. 1. Gasworks. 1895. 2. Southdown Brewery Company. 1896.

Made and communicated by MR. G. BATES.

Good supply of water in both.

	Feet. (1)	Feet. (2)
Mixed red earth	10	10
Blue marl and alluvial deposit	40	40
Chalk and flints	15	32
Totals	65	82

LEWES. Phœnix Ironworks. About marsh-level.

Communicated by MESSRS. WELLS & Co.

[Alluvium] Clay, 50 feet.

LEWES. Close to the river.

Made and communicated by MESSRS. WELLS & Co.

A driven tube of 2 in. diameter.

River-deposits, to Chalk, 30 feet.

LEWES. Waterworks. At the edge of the marsh, close to strong springs.

Communicated by the Company.

Sunk 24 ft., the rest bored.

[Drift] { Red gravelly loam 20 }
{ Chalk-gravel 4 } 144 feet.
[Upper] Chalk, with flints 120 }

For Analysis of the water see pp. 114, 115.

LITTLEHAMPTON. Waterworks. 1888.
Communicated by MR. R. F. GRANTHAM.
About 24 feet above Ordnance Datum.

First shaft about 60 feet, the rest bored. Second shaft 80 feet, with galleries from the bottom, east and west, for 76 yards.

Water found in the Chalk with flints, in the galleries (none in the boring). 168,000 gallons pumped in 24 hours.

		Thickness.	Depth.
		Feet.	Feet.
[Drift, 19½ feet]	Brickearth	7	7
	Earth and sand	5	12
	Stiff clay and sand	7½	19½
[Upper Chalk, 95½ feet]	Chalk, dyed yellow	5½	25
	Pervious white chalk	17	42
	Hard white chalk	12	54
	Hard white chalk with a few flints	5	59
	Hard white chalk with many flints	27	86
	Solid white chalk, very hard, no flints	29	115
[Middle Chalk and Lower Chalk, 391 feet]	Impervious clunch	8	123
	Hard white chalk	236	359
	Clunch	2	361
	Blue chalk marl, very hard	6	367
	[Undescribed]	11	378
	Soft chalk, light blue	35	413
	Solid white chalk	61	474
	Impervious grey chalk	32	506

If the classification suggested in square brackets be approximately correct, we might expect to reach Chloritic Marl and Upper Greensand within a few feet, for the combined thickness of the flintless Lower and Middle Chalk in Sussex is usually about 400 feet. In the absence of specimens it is impossible, however, to identify the different zones.

According to Dr. Kelly's Report for 1887, a shaft of 6 feet diameter was carried to the depth of 60 feet, then one of 3 feet diameter for 9 feet, then a boring of 9 inches diameter for 150 feet, and then one of 8 inches to 358 feet (at the end of 1877).

LITTLEHAMPTON. Anchor Brewery. About 1830 (or soon after).
Communicated by MR. T. CONSTABLE (partly from a letter by Mr. W. Dyer, the former owner).

Bored throughout (there is also a well of 20 feet, about 12 feet off).

Water-level 6 feet down, not decreased after pumping 12 hours. Has always been the same.

	Thickness.	Depth.
	Feet.	Feet.
Sandy loam	5	5
Hard chalk, with layers of flints (water found 12 feet down in this) - - - about	95	100
What appeared to be a very stiff pipe-clay, but burnt to lime [soft chalk]	211	311
Undescribed	2	313

Boring not carried deeper because the rods were too slight. Plenty of water for the first 100 feet, but none after.

When the channel of the brook, some 400 yards off, nearer the sea, was cleared, on cutting through the clay they came into marl [chalk], and the water in the well then became salt. However, after that part of the drain that passed through the marl was puddled with clay, the saltness gradually decreased until it disappeared.

For Analysis of the water see p. 115.

LITTLE HORSTED. Wicklands, at the bend of the road, three-quarters of a mile S.W. of the village.

120 feet above Ordnance Datum.

Shaft, with 35 feet of water from the Lower Tunbridge Wells Sand.

[?Upper Tunbridge Wells Sand.] Marly clay and sandstone 35
[?Grinstead Clay.] Blue clay - - - - - - - 6 } 50 feet.
[?Lower Tunbridge Wells Sand.] Sand (and sandstone?) - 9

According to information from MR. J. LUCAS, an old well at a cottage 400 feet S.W. of the above, and 115 feet above Ordnance Datum, is 50 feet deep in sand, with 6½ feet of water. Marl not having been reached, the water comes from the Upper Tunbridge Wells Sand. The water is low in autumn.

LODSWORTH. Messrs. Tallants, for Earl of Egmont. 1883.

Made and communicated by MESSRS. LE GRAND & SUTCLIFF.

		Thickness.	Depth.
		Feet.	Feet.
[Lower Greensand]	Brown sandy soil	6	6
	Sandy clay	10	16
	Dark sand, with loose shale	29	45
	Loose shaly sandstone	2	47
	Hard sandstone	2	49

LOXWOOD. Tichbourne Public House. 1889?

Made and communicated by MESSRS. DUKE & OCKENDEN.

		Thickness.	Depth.
		Feet.	Feet
[Weald Clay]	Clay. At 228 feet down, changes from deep red to dark blue	380	} 430
	Hard rock [Paludina-marble] a few inches.		
	Clay	30	

A quarter of a mile south, at Loxwood House, a good supply was got in marly clay at 31 feet.

MADEHURST Dalepark.

Made and communicated by MESSRS. A. WILLIAMS & Co.

Water-level varies from 150 to 450 feet down.

Shaft, in Chalk - - - - 320 } 470 feet.
Boring, in Chalk with flints - 150

MAYFIELD. Convent (the Old Palace). About 300 yards N. of the building, and at a level about 50 feet lower.

From letters from the Mother Superior, with details of the beds from the borer, MR. HYMAS.

The spring met with at the depth of about 50 feet was inadequate, and this supply escaped at two lower depths. The yield was tested at the depth of 105 feet and found to be at the rate of 30 gallons an hour. Unsuccessful.

	Thickness.	Depth.
	Feet.	Feet.
Shaft (the rest bored)	—	24
Blue clay	4	28
White sand-rock	8	36
Blue clay and sand	8	44
Blue clay. Water found	6	50
Blue shale	8	58
Brown rock	2	60
Yellow clay and sand	8	68
Sand-rock	6	74
Blue clay	4	78
Clay sand	9	87
Brown rock	3	90
Blue shale	8	98
Blue clay	3	101
Sand-rock. Water lowered	6	107
Blue shale	14	121
Shale and stone	14	135
Sand-rock	20	155
Blue shale	4	159
Sand-rock	6	165
Clay and sand	3	168
Shale and stone. Water lowered	8	176
Sand-rock	10	186
Sand shale	2	188
Sand-rock	12	200
Clay sand	4	204
Soft sand-rock	10	214
Soft shale	10	224

Another well. Made and communicated by MESSRS. ISLER & Co. 1895. Water-level 193 feet down. Supply abundant.

		Thickness.	Depth.
		Feet.	Feet.
Sunk Well	[Clay, unevenly on the bed below. (Note of MR. TOPLEY'S)]	98	98
	Hard sandstone	$9\frac{1}{2}$	$107\frac{1}{2}$
	Hard rock	$12\frac{1}{2}$	120
	Sandstone	$70\frac{3}{4}$	$190\frac{3}{4}$
	Sandy shale	$8\frac{1}{4}$	199
	Sandstone	$2\frac{1}{4}$	$201\frac{1}{4}$
	Light [-coloured] shaly marl	3	$204\frac{1}{4}$
	Sandstone	$4\frac{1}{2}$	$208\frac{1}{2}$
	Shale	$1\frac{1}{2}$	210
	Marl	$59\frac{1}{2}$	$269\frac{1}{2}$

MERSTON. Trial boring for Bognor Waterworks.

Made and communicated by MESSRS. DOCWRA.

? Water-level about 10 feet down.

		Thickness.	Depth.
		Ft. in.	Ft. in.
Made ground	- - - - - -	1 0	1 0
[Drift, 12½ feet]	Light ballast - - - -	9 0	10 0
	Light sand - - - -	2 0	12 0
	Clay and ballast - - -	1 6	13 6
	Hard blue clay - - -	4 0	17 6
	Soft clay with sand - -	12 6	30 0
	Blue clay - - - -	16 0	46 0
	Sandy clay - - - -	2 0	48 0
[London Clay, 292½ feet]	Blue clay - - - -	95 6	143 6
	Green sand - - - -	1 6	145 0
	Hard rock - - - -	0 10	145 10
	Blue clay - - - -	88 2	234 0
	Hard rock - - - -	1 0	235 0
	Hard blue clay - - -	65 0	300 0
	Blue clay - - - -	6 0	306 0
[Reading Beds, 99½ feet]	Mottled clay - - - -	8 0	314 0
	Hard red clay - - -	91 0	405 0
	Flints - - - -	0 6	405 6
	Chalk marl - - - -	4 6	410 0
[Upper Chalk, 244½ feet]	Chalk, with six inches of flints at the base - -	45 6	455 6
	Chalk and flints - - -	59 6	515 0
	Hard chalk - - - -	8 0	523 0
	Chalk and flints - - -	27 0	550 0
	Hard chalk - - - -	94 0	644 0
	Mild chalk - - - -	6 0	650 0

MIDHURST. Rev. H. Back's, Ashfield, opposite Gulland's Oak (Gilders' Oak F. of the old Ordnance Map). 1885.

Sunk and communicated by MESSRS. LE GRAND & SUTCLIFF.

Water-level **46 feet** down. Good supply.

		Thickness.	Depth.
		Feet.	Feet.
[Folkestone Beds]	Sandy brown clay - - -	10½	10½
	Ironstone - - - -	1	11½
	Sandy loam - - -	11	22½
	Dark sandy clay - - -	3	25½
[Sandgate Beds]	Light-grey sand - - -	19	44½
	Yellow sand - - - -	7½	52
	Dark-green clayey sand - -	8½	60½
	Light-green sand - - -	14½	75
	Dark-green sandy clay - -	4	79
[Hythe Beds]	Dark dead sand - - -	2	81
	Yellow sandstone - - -	19	100

A letter from MR. BACK makes the top part "mixed gravel and sand and a little clay, about 12 feet, then one foot of ironstone rock" (which he thinks is probably more correct), and notes that there was a little water at a depth of 55 feet, but more at 79 feet (rather ferruginous).

MIDHURST. Pitsham Brickfield, about a mile S.W. of the town. Messrs. Tallant. For Lord Egmont. 1883.
Made and communicated by MESSRS. LE GRAND & SUTCLIFF.
Water-level 28 feet down.

		Thickness.	Depth.
		Feet.	*Feet.*
[Folkestone Beds]	Old dug well (the rest bored)	—	15
	Variously coloured hard sand, with ironstone	31	46
	Bands of white clay and yellow sand	2	48
[Sandgate Beds]	Clay	1	49

MIDHURST. For other wells, *see* EASEBOURN.

MID-SUSSEX WATERWORKS, *see* BALCOMBE.

MOUNTFIELD. "The Sub-Wealden Exploration."

About 60 feet eastward of the bed of the stream separating Councillor's Wood from Lime Kiln Wood, according to "Sub-Wealden Explorations, First Quarterly Report."

"The Geology of the Weald," by W. TOPLEY, 1875, pp. 42-49 (and other sources).

First Boring, finished 1874. 9 inches diameter to 312 feet, then 4 inches to 328 feet, the rest 3 inches.

		Thickness.	Depth.
		Feet.	*Feet.*
Purbeck Beds	Shales	$16\frac{1}{2}$	$16\frac{1}{2}$
	Blue limestone, with spring	$2\frac{1}{2}$	19
	Shale	5	24
	Blue limestone	2	26
	Shale	4	30
	Limestone	$1\frac{1}{2}$	$31\frac{1}{2}$
	Shale	4	$35\frac{1}{2}$
	Limestone	3	$38\frac{1}{2}$
	Shale, with spring. Water stood permanently at 42 feet down, inside the tubes	4	$42\frac{1}{2}$
	Limestone	4	$46\frac{1}{2}$
	Hard blue shale	$15\frac{1}{2}$	62
	Hard grey shale	3	65
	Hard shale	$14\frac{1}{2}$	$79\frac{1}{2}$
	Shales, with crystals of carbonate of lime	9	$88\frac{1}{2}$

MOUNTFIELD. "The Sub-Wealden Exploration"—continued.

		Thickness.	Depth.
		Feet.	Feet.
Purbeck Beds	Grey shale	13	101½
	Greenish shales, with gypseous veins	20	121½
	Impure gypsum	8½	130
	Pure white gypsum (alabaster)	4	134
	Impure gypsum	5½	139½
	Pure white gypsum (alabaster)	3	142½
	More or less pure, hard and dark gypsum	14½	157
	Black shale, very sulphureous	3½	160½
	Gypsum in nodules and veins	12	172½
	Gypseous marl	6½	179
	Sandy marl. Water-level lowered here	½	179½
	Black sulphureous shale	½	180
? Portland Beds, 110 feet	Greenish sand, with nodules of black chert	21	201
	Sandy shale	30	231
	Calcareous matter, with chert-nodules	8	239
	(Not described)	2	241
	Hard black sandy shale, very sulphureous	12	253
	Blacker and softer shale	7	260
	Harder shale, with much chert	12	272
	Black shale, very sulphureous	14	286
	Paler shale, with veins of gypsum	4	290
	Darker and more sandy shale	2	292
	Shale	2	294
	Dark clay	18	312
Kimeridge Clay, 727 feet	Clay, generally rather sandy, some calcareous (toward the lower part)	288	600
	Hard light-coloured bed, very rich in petroleum	2	602
	Clay, with bands of cement-stone	232	834
	Cement-stone	50	884
	Clay	2½	886½
	Cement-stone	2½	889
	Clay	67	956
	Dark clay, with cement-stone	55	1011
	Sandy bed	2	1013
	Dark clay	4	1017

The lowest 61 feet were originally classed as Oxford Clay; but the second boring showed that the Kimeridge Clay goes much deeper and is succeeded by Corallian Beds.

A core of some 17 feet, or to the depth of about 1,030 feet was left in the borehole. The work was stopped by an accident to the rods.

A list of the fossils found, from 300 to 1,013 feet down, is given in the "Wealden Memoir," p. 44.

No complete section of this boring is given in the "Quarterly Reports of the Exploration;" but in the second of these, some details from 131 feet downward, differ from the above account.

White gypsum (alabaster) reached at 131, 4 feet thick, or to depth of 135
Gypseous marl - - - - - 10 ,, ,, 145
Alabaster - - - - - 3 ,, ,, 148

MOUNTFIELD.

"The Sub-Wealden Exploration." Second Boring. Begun February 1875, finished 1876.

Tenth and Twelfth Quarterly Reports, in the "Record of the Sub-Wealden Exploration," by H. WILLETT. 8vo, Brighton, 1878. This also gives the amount of core brought up, and the amount done each day, down to 1,546 feet. These details are given by MR. THORNTON. Those below 1,546 feet are from a lithographed section issued by the Aqueous Works and Diamond Rock-Boring Co. Boring of 8 inches diameter at first, decreasing to 2 inches at last. Some further details from an account by W. TOPLEY, *Rep. Brit. Assoc.* for 1880, p. 105.

		Thickness.		Depth.	
		Ft.	in.	Ft.	in.
	Alluvial deposit	—	—	16	0
	Soft shale	1	0	17	0
	Blue limestone	1	6	18	6
	Calcareous shale	6	0	24	6
	Blue limestone	1	0	25	6
	Calcareous shale	1	0	26	6
	Soft shale	3	0	29	6
	Limestone	1	6	31	0
	Calcareous shale	0	6	31	6
	Strong shale	3	0	34	6
	Calcareous shale	1	0	35	6
	Blue limestone	2	6	38	0
	Calcareous shale	1	0	39	0
	Strong shale	6	0	45	0
	Blue limestone	2	0	47	0
	Shale	0	6	47	6
	Hard limestone	1	6	49	0
	Limestone and soft shale	8	0	57	0
	Shale	5	0	62	0
	Calcareous shale	1	0	63	0
	Shale	7	0	70	0
	Blue limestone	1	0	71	0
	Strong shale	6	3	77	3
	Compact blue limestone	1	3	78	6
	Strong shale	6	9	85	3
	Calcareous shale	2	6	87	9
[Purbeck Beds]	Strong shale, with limestone at 93 ft. 11 in. to 94 ft.	11	3	99	0
	Compact hard shale	7	0	106	0
	Calcareous shale	2	4	108	4
	Hard limestone	0	8	109	0
	Hard blue shale	6	6	115	6
	Blue limestone	1	0	116	6
	Dark blue shale	6	6	123	0
	Shaly limestone, with thin veins of broken gypsum	4	0	127	0
	Impure gypsum	6	0	133	0
	Limestone and gypsum (thin veins)	1	0	134	0
	Shaly gypsum	2	3	136	3
	Gypsum in crystals, veins in shale	7	6	143	9
	Gypsum in veins and nodules	3	6	147	3
	Gypsum, with veins of limestone	1	0	148	3
	Strong shale and veins of limestone, with gypsum	0	6	148	9
	Strong shale, with nodules of gypsum	4	3	153	0
	Gypsum, more or less pure	7	9	160	9
	Strong shale, with gypsum	3	3	164	0
	Nearly pure gypsum, with veins of carbonate of lime	4	4	168	4

Mountfield. "The Sub-Wealden Exploration." Second Boring—cont.

		Thickness.		Depth.	
		Ft.	in.	Ft.	in.
[Portland Beds, 105¾ ft.]	Fragments of shale and chert. Water ran away at 169 ft. and tool dropped 4 ft., and again lower down - - - -	24	4	192	8
	Soft sandy shale - - -	7	4	200	0
	Soft whitish sandstone - -	52	0	252	0
	Soft sandstone, darker - -	5	0	257	0
	Sandy shale - - - -	17	0	274	0
	Kimeridge clay - - -	109	0	383	0
	Kimeridge clay, rather softer -	45	0	428	0
	Kimeridge clay, more compact	44	0	472	0
	Kimeridge clay, softer - -	23	0	495	0
	Kimeridge clay, solid - -	26	0	521	0
	Kimeridge clay, with traces of carbonate of lime - -	20	0	541	0
	Dark brown Kimeridge clay -	66	0	607	0
	Brown limestone - - -	1	6	608	6
	Kimeridge clay - - -	3	0	611	6
	Brown limestone - - -	0	6	612	0
	Kimeridge clay - - -	27	0	639	0
	Kimeridge clay, with veins of carbonate of lime - -	40	0	679	0
	Kimeridge clay, very calcareous	21	0	700	0
	Kimeridge clay, much softer and darker, very full of fossils - - - - -	24	0	724	0
	Kimeridge clay, with large veins of carbonate of lime -	17	0	741	0
[Kimeridge Clay, †1290 ft.]	Kimeridge clay, with smaller veins of carbonate of lime -	22	0	763	0
	Kimeridge clay, with small veins of carbonate of lime -	18	0	781	0
	Kimeridge clay - - -	19	0	800	0
	Kimeridge clay, with veins of carbonate of lime - -	10	0	810	0
	Kimeridge clay - - -	72	0	882	0
	Kimeridge clay, with hard bands of limestone. A very soft place at 922 ft. - - -	57	0	939	0
	Clay - - - - -	16	0	955	0
	Clay, with veins of carbonate of lime - - - - -	28	0	983	0
	Oxford clay, harder and more calcareous - - -	9	0	992	0
	Oxford clay, more sandy and very soft, with veins of carbonate of lime - - -	12	0	1004	0
	Sandstone, rather shaly and full of fossils - - -	41	0	1045	0
	Sandy shale - - - -	2	0	1047	0
	Sandy shale, more compact and solid - - - -	17	0	1064	0
	Sandy shale, with nodules of limestone - - - -	28	0	1092	0
	Shaly sandstone - - -	16	0	1108	0
	Very shaly sandstone - -	21	0	1129	0

MOUNTFIELD. "The Sub-Wealden Exploration." Second Boring—*cont.*

		Thickness.		Depth.	
		Ft.	in.	Ft.	in.
	(Undescribed). All the core left in the hole. [Sandstone, very shaly, TOPLEY]	8	0	1137	0
	Shaly limestone	27	0	1164	0
	Light-blue limestone	4	0	1168	0
	Shaly limestone	14	0	1182	0
	Calcareous shale	28	0	1210	0
	Calcareous shale, more free from sand	26	0	1236	0
	Very clayey shale, more like Oxford Clay	19	0	1255	0
	Calcareous shale	21	6	1276	6
	Soft dark gritty limestone	28	6	1305	0
	Calcareous shale	20	0	1325	0
[Kimeridge Clay, ?1290 ft.]	Friable calcareous grit	17	0	1342	0
	Soft calcareous grit, with bands of hard limestone	24	0	1366	0
	Limestone	4	0	1370	0
	Blue limestone changing into shale	27	0	1397	0
	Strong blue shale, with few fossils	19	0	1416	0
	Strong blue shale	4	0	1420	0
	Limestone, very full of oyster-shells	10	0	1430	0
	Blue very calcareous shale	16	0	1446	0
	Shale, with very few fossils	20	0	1466	0
	Blue shale, few fossils for 11 ft., then traces of encrinites	60	0	1526	0
	Blue shale, with a great many encrinites and other fossils	38	0	1564	0
	Calcareous shale, with hard bands of limestone	88	0	1652	0
	Light-blue limestone	10	0	1662	0
	Calcareous shale and fossils	9	0	1671	0
[? Corallian, 222 ft.]	Calcareous shale, with hard limestone	27	0	1698	0
	Very soft dark shale, with a great many fossils	59	0	1757	0
	Strong dark shale	12	0	1769	0
	Hard grey limestone	17	0	1786	0
	Dark sandy shale	26	0	1812	0
[Oxford Clay, 120 ft. ?]	Dark shale	12	6	1824	6
	Shale	81	6	1906	0

The classification is taken, as nearly as can be, from that of H. B. WOODWARD in the "Memoir on the Jurassic Rocks of Britain," Vol. v., pp. 346, 347 (1895). But his account of the Purbeck and Portland Beds does not tally with the above details, whereas it does agree much more with those of the first boring. There is no doubt that in various accounts the two borings have been rather mixed up, and that some error has crept in by reason of this.

Mr. Woodward's classification is as follows, with the figures given above on the left :—

Feet.				Feet.	
274 { 168½	Purbeck Beds	-	-	177 } 292,	whereas Kimeridge Clay
105⅔	Portland Beds	-	-	115 }	clearly begins after 274.
? 1290	Kimeridge Beds	-	-	1273	
? 241	Corallian Beds	-	-	241	
? 120	Oxfordian Beds	-	-	99	
				1905	

His details, too, differ from the above, but are not so full. I must own to some doubt as to the classification.—W. W.

For Newhaven and Seaford Waterworks, see EAST BLATCHINGTON.

NEWICK. For Dr. Hughes. 1898 ?

Boring made and communicated by MESSRS. ISLER.

Water-level 12 feet down. Supply 360 gallons an hour.

		Thickness.	Depth.
		Feet.	Feet.
Well [? old]	-	—	70
	Blue marl	24	94
	Grey sand	2	96
	Sandstone	1	97
	Blue marl	11	108
	Sand-rock	3½	111½
	Blue clay	5½	117
	Sand-rock	3	120
	Sandstone and marl	5	125
	Blue marl	1	126
	Brown marl	1	127
	Mottled clay	14½	141½
	Blue marl	1	142½
	Blue rock	1	143½

NEWICK. Cobb's Nest (? half-a-mile northward of Parsonage.)

H. W. BRISTOW, in "The Geology of the Weald," p. 88. 1875.

Water came in on the northern side, on top of the sand-rock.

	Thickness.	Depth.
	Feet.	Feet.
? What (? sand in part), - - - about	41	41
? Grinstead Clay. Tea-green and purple variegated shale, the lower part harder and more gritty, - - - - - - - about	20	61
Rock - - - - - - -	3	64

NEW TIMBER (near).

DR. MANTELL. "The Fossils of the South Downs," 4to, Lond., 1822, p. 84.

Gault { Grey chalk marl, gradually passing down into the next 20 } 90 feet
      { Blue chalk marl, with many *Ammonites, Inocerami*, &c. 70 }

NORTH MUNDHAM. Runcton House.

Bored and communicated by MESSRS. DUKE & OCKENDEN.

Good supply of water, standing within 3 feet of the surface.

		Thickness.	Depth.
		Feet.	*Feet.*
[Reading Beds]	Old dug well	—	20
	Clay	60	80
[Upper] Chalk	with black flints	45	125

NORTH MUNDHAM. The Vicarage.

Abundance of water. Communicated by MR. OCKENDEN.

		Thickness.	Depth.
		Feet.	*Feet.*
[Drift]	Sand	6	6
[Reading Beds]	Red clay, at 18 feet		
	Mottled clay, at 46 feet		
	Rock (9-inch), at 66 feet	69	75
	Rock, at 70 feet		
	Rock, at 72½ feet		
[Upper Chalk]	Chalk and flints, with pink clay at bottom	82	157

NUTHURST. Manning's Heath. Close to the Dun Horse.

Bored and communicated by MESSRS. DUKE & OCKENDEN.

Good supply at the depth of 48 feet.

		Thickness.	Depth.
		Feet.	*Feet.*
	Old well (the rest bored)	—	32
[? Tunbridge Wells Sand]	Rock	2	34
	Sand	3	37
	Blue rock	44	81

At NUTHURST Lodge a well was sunk 80 feet (belled out from 60 feet) and then a boring was made for 27 feet. Narrow fissures were cut at 67 feet, running W. or N. of W. The water-level is 75 feet down. There is an older well here.

PAGHAM. Sefter School.

From samples communicated by MESSRS. DUKE & OCKENDEN, who sunk the well.

		Thickness.	Depth.
		Feet.	Feet.
[Drift 18 feet]	Loamy gravel	8	8
	Shrave	10	18
[LondonClay, 157 feet]	Hard blue clay, more or less sandy	6	24
	Buff and brown sand	8	32
	Buff sand	14	46
	Blue clay and shells	11	57
	Brownish and blue clay	11	68
	Blue clay, not sandy	23	91
	Blue clay, more soapy	9	100
	Blue clay, with pyrites and fragment of large oyster	9	109
	Blue and brownish clay	34	143
	Blue and brownish clay, more sandy	13	156
	Blue and brownish clay, still more sandy	11	167
	Black sandy loam	5	172
	Brown sandy loam	3	175
[Reading Beds, 104 feet]	Mottled red and brown clay and sand	7	182
	Grey loam	7	189
	Blackish loam	3	192
	Mottled clays	87	279
[Upper Chalk]	Chalk and flints (no water)	188	467

Heavy charges of dynamite were exploded in this well, to increase the yield of water, but without result, and the well has been abandoned.

PATCHAM. Brighton Waterworks. Third Pumping Station, less than half a mile westward of the church. 1886. Galleries extended later.

Information from MR. J. JOHNSTON.

Ground-level at the engine-house 195·2 feet above Ordnance Datum. The bottoms of the headings 174½ feet lower.

The wells are elliptical, longer diameter 12 feet, and shorter diameter 8 feet. The directions of the chief headings approximately N.E. and S.W., with a shorter one S., for about 410 feet. Total length 1,727 feet, but being extended.

Average daily yield in 1895, 1,200,000 gallons.

The following notes on the galleries here were made in September, 1893, from personal inspection (W. W.). They were all in firm chalk.

The western gallery then reached to 125 feet from the pumping-shaft and showed a marked continuous layer of flint. Practically no water found till reaching the end, where there was a good spring along a small fault (? 9 inches throw). The beds mostly flat, but the flint-layer sometimes queerly broken.

The eastern gallery, from the pumping-shaft to another shaft (Robey Engine) had practically no water. At the Robey shaft water is said to come in, some way up, after heavy rain, showing ready communication with the surface. Further on was another case of like communication down a fissure from the surface, the gallery has given way at the top and water is said to come in 24 hours after rain. Still further there was a good spring at the bottom of the channel along the bottom, forming a hole. Apparently the beds rise slightly eastward and the marked flint-layer is lost very soon after leaving the pumping-shaft

The southern gallery had hardly any water till getting to the end, 230 feet from the Robey shaft, where there was a small spring. Just here the roof had given way on account of rotten flint beds at the top, which had therefore been narrowed; elsewhere the galleries have a nearly flat roof, sometimes over 7 feet wide.

PATCHING. Cottage close to house for Mr. Goad.
Made and communicated by MESSRS. DUKE & OCKENDEN.
Water stands 8 feet down.

		Thickness.	Depth.
		Feet.	Feet.
[Reading Beds]	Blue and yellow clay	30	30
	Blue and black clay	25	55
	Clay and flint	10	65
	Clay	2	67
[Upper Chalk]	Hard flint and chalk	58	125

PETWORTH House.

P. J. MARTIN. "A Geological Memoir on a Part of Western Sussex," p. 36, 4to, Lond., 1828, and W. TOPLEY, note in "The Geology of the Weald," p. 116, 1875.

		Thickness.	Depth.
		Feet.	Feet.
Hythe Beds	Loose sandy rock	16	16
	Whin (? chert)	2	18
	Sandstone, sand and whin (? chert)	35	53
	Whin (? chert)	5	58
	Rocky sand. Water	7	65
Atherfield Beds	Black sand	35	100
	Brown sand. Water	15	115
Weald Clay	Clay	281	396
	Pyrites	1	397
	Greenish-grey sand	3	400

PEVENSEY SLUICE. House marked on the old Ordnance Map (Sheet 5) northward of Martello Towers 52, 53.
A tube-well struck rock at the depth of 20 feet and got salt water.

PLASHETT PARK. Near a cottage in the south-eastern corner.
DR. MANTELL. "The Fossils of the South Downs," 4to, Lond., 1822, p. 66.

		Thickness.		Depth.	
		Ft.	in.	Ft.	in.
Weald Clay	Ochraceous loam	5	0	5	0
	Weald clay	5	0	10	0
	Sussex marble	0	5	10	5
	Weald clay	5	0	15	5
	Sussex marble	0	10	16	3
	Weald clay	9	0	25	3
	Sussex marble. To spring of excellent water	0	10	26	1

POLEGATE, see HAILSHAM.

PORTSLADE. Aldrington Waterworks, a quarter of a mile north of the Station. Since acquired by the Corporation of Brighton.

Boring made and communicated by MESSRS. LE GRAND & SUTCLIFF.

Water-level 65 feet down.

		Thickness.	Depth.
		Feet.	Feet.
Shaft	[? Drift and Chalk]	—	74
[Upper Chalk]	Hard chalk and flints	28	102
	Chalk and flints, free cutting	33	135
	Hard chalk and flints, free cutting from 201 to 213	169	304

PORTSLADE. Brewery (Mew's). 1884.

Made and communicated by MESSRS. DOCWRA.

Shaft, with gallery (base 10 feet above bottom of shaft). Water-level, 57½ feet down.

	Thickness.	Depth.
	Feet.	Feet.
Made ground	5	5
Comb rock	18	23
Chalk and flints	65½	88½
? bored deeper		

PORTSLADE. Brickyard, southern side of Brighton Road, about 15 feet below old surface level, and about 60 feet above Ordnance Datum.

		Thickness.	Depth.
		Feet.	Feet.
[Woolwich and Reading Beds, perhaps reconstructed]	Brick earth	9	9
	Sand and clay	11	20
	Sand with flints	20	40
[Upper Chalk]	Chalk. To water	4	44

PULBOROUGH. Borough Farm. 1898.
Bored and communicated by MESSRS. DUKE & OCKENDEN.
Water-level at 31 feet from the surface. Supply small.

		Thickness.	Depth.
		Feet.	*Feet.*
	Soil and loam	4	4
[Hythe Beds?]	Sand-rock, loose stones at about 10 to 12 feet	8	12
	Sand and stones	8	20
	Yellow clay mixed with sand	4	24
[Atherfield Clay?]	Green and yellow sand with clay	2	26
	Sand and clay	5	31
	Wet sand (about the level of old dug well)	2	33
[Weald Clay]	Yellow sand and clay	12	45
	Blue clay	7	52
	Brown [and purple] clay	5	57

PULBOROUGH.

P. J. MARTIN. "Geological Memoir on a Part of Western Sussex,
p. 30, 1828.

Gault? { Well (the rest bored) - 30 }
{ Sandy blue clay, to sand rock, with a copious supply } 65 feet.
{ of good water, which rose 18 ft. above the boring 35 }

RINGMER. Five wells from Dr. Mantell's "Fossils of the South Downs."
Quarto, London, 1822, pp. 75, 82, 83.

Park-house. Chlorite [glauconite] sand, 40 feet.

Moor Lane. Cottage. Blue marl, the lower beds with much green sand and some fossils (Gault), 50 feet.

Norlington Green. (Gault.) Blue marl, with very many shells below 15 feet. At 20 feet a layer of red marl, a few inches thick, and another 10 feet lower. 50 feet.

Cottage near Mr. W. Green's house :—
A spring of excellent water suddenly appeared at the bottom, and the water rose 10 feet.

		Thickness.	Depth.
		Feet.	*Feet.*
	Yellow ochraceous loam	5	5
	Blue marl, with *Ammonites, Inocerami, Hamites,* and selenite	15	20
Gault	Dark blue marl, inclining to black. Small crystals of selenite in the upper part, and in the lower nodular masses of hard marl, with green sand, quartz grains, and pyrites	10	30
	Green chlorite [glauconite] sand	4	34

RINGMER. Mr. W. F. Martin's.

From samples communicated by MR. MARTIN.

[Lower Greensand ?]
- Clay and nodule, at 143 feet.
- Hard micaceous clay (bottom of sunk well) at 150 feet.
- Blackish clay and green sand at 160 feet, more sandy at 165 and 180 feet.
- Green sand at 188 feet.

No record has been kept of the old sunk well, which was probably entirely in Gault. Perhaps the beds to 180 feet belong to the Gault.

RINGMER. Public well, on the Green. 1883. 72 feet above Ordnance Datum.

From samples (taken occasionally) communicated by MR. W. F. MARTIN, of Ringmer.

[Gault, 130 feet]
- Soil.
- Weathered clay, at 2 feet.
- Grey clay, at 19 feet; *Ammonites*, at 25 feet; fossils, at 30 feet; *Dentalium*, at 36 feet; shelly, at 44 and 50 feet; *Inoceramus*, at 60 and 70 ft.
- Ochre, at 74 feet.
- Grey shelly clay, at 80 feet; and cement-stone, with *Nucula pectinata*, at 85 and 90 feet; Ammonite, at 96 feet.
- Grey shelly clay, at 108 and 110 feet.
- Hard bed, with *Inoceramus* and phosphatic nodules at 120 to 130 feet.

[Lower Greensand ?]
- Greenish sandy clay, at 170 feet.
- Coarse quartz-sand, with small quartz pebbles, and glauconite (good water), 190 to 218 feet.

Compare with West Firle (p. 97), four miles to the south-east, where the "greenish sandy clay" rests directly on Weald Clay.

Another version, communicated by MR. G. FULLER, of Lewes, adds, the shaft is 150 feet deep, the rest being bored, and that the water-level is 39 feet down. It gives a different classification, as follows:—

	Thickness.	Depth.
	Feet.	*Feet.*
Gault	189	189
Mixture of gault and sand. Pipe continually choked by hard lumps of a conglomerate of gault and sand. Thin layer of shale - -	19	208
Lower Greensand - - - - - - -	6½	214½

ROTHERFIELD. Maynard's Gate. For the Crowborough District Water Company.
Made and communicated by Mr. A. E. Nunn, and from Mr. C. O. Blaber.
322 feet above Ordnance Datum.

Shaft, of 10 ft. diameter, to 56 feet ; the rest a bore of a foot diameter.

			Thickness.	Depth.
			Feet.	*Feet.*
Soil	-	-	2	2
[Drift]	Gravel - - -		3	5
	Dark clay - - -		24	29
	Brown sand-rock. Dip 65° [?] to N.E. On reaching this, the yield was at the rate of 60,000 gallons of water in 24 hours - - - -		8	37
	Hard stone, on piercing which the supply increased to 150,000 gallons a day - - - -		1	38
	Blue clay - - -		3½	41½
Ashdown Sand	White sand-rock, dip 30° [?] to N.E. - -		3	44½
	Blue clay - - -		1¼	45¾
	White rock, on reaching which the supply increased to 216,000 gallons a day -		17¼	63
	Soft clay - - -		3	66
	Hard blue clay - - -		17	83
	Sand-rock - - -		6	89
	Clay - - - -		4	93
	Sand-rock. Supply 303,000 gallons a day - -		2	95

[The recorded dips probably represent current-bedding in the sand. Measurements from 5 to 45¾ feet taken on the side highest by dip.]

RUDGWICK. Hermongers. For Mr. T. T. Busk. 1890.

Communicated by Mr. Busk.

Shaft about 55 feet, the rest bored.

The supply, from the bottom did not fail (from August, 1890, to February, 1891), although many wells in the parish were dry during the long frost. Water very hard, but otherwise satisfactory. It rises up the bore and stands at about 43 feet down the well.

[Weald Clay]	Blue and red clay. A little water at the depth of 18 feet - - - - - Clay - - - - -	Varying in thickness and reaching nearly to the base of the shaft on one side.
	Brown rock, in the bottom of the shaft and in the top of the bore. Grey rock. Clay.	

RUDGWICK. Upper **Hillhouse Farm**, nearly two miles westward of the village. 1891?

From specimens sent by MESSRS. TILLEY and from workman's note-book.
Boring of six inches diameter.

Water at first stood 15½ **feet down.** Started falling when the boring was 60½ feet deep, to 19 feet, and when the boring was 80 feet deep stood at 52 feet.

		Thickness.	Depth.
		Feet.	*Feet.*
[Weald Clay]	Sand, dry at top	15	15
	Red clay	5	20
	Bluish sandy clay	4	24
	Brown clay	10	34
	Bluish clay	6	40
	Yellow sandy clay	5	45
	Stiff yellow clay	5	50
	Blue clay	17	67
	Sandstone	½	67½
	Blue clay	32½	100

RYE. Batchelor's Brewery. Northern end of the town, south of the railway.

About 15 to 20 feet above Ordnance Datum.

From MR. J. ELLIOTT, in "Geology of the Weald," p. 49. 1875.

		Thickness.	Depth.
		Feet.	*Feet.*
Alluvium	Clay	3 or 4	10?
	Peat, with logs of wood	6 or 8	
Fairlight Clays	White and red mottled, with several layers of sandstone (one, at a depth of about 150 feet, 23 feet thick), and a few thin layers of hard rock	330	340?

RYE. Under Cadborough Cliff, 1¼ miles from the town. For public supply. 1898?

Boring communicated by MR. P. H. PALMER.

Yield 160,000 gallons a day, not lowering the water more than 9 feet below the ground. After half an hour's cessation of pumping the water overflows, a foot above the ground, at the rate of 40,000 gallons in twenty-four hours. Water very good.

		Thickness.	Depth.
		Feet.	*Feet.*
[Alluvium]	Peat	18	18
	Shale	14	32
	Sand-rock	30	62

SEAFORD—SELSEY.   79

SEAFORD. Waterworks, *see* EAST BLATCHINGTON.
SEFTER, see PAGHAM.

SELMESTON. Mr. C. Long's Cottages, by Reading Room, S. of Church. 1888.
Made and communicated by MESSRS. LE GRAND & SUTCLIFF.
Water-level 50 feet down.

			Thickness.	Depth.
			Ft. in.	*Ft. in.*
Top soil			0 6	0 6
	Yellow Clay		5 6	6 0
	Blue clay		12 0	18 0
[Gault]	Red shale		0 2	18 2
	Mixed coloured clay and sand		15 0	33 2
	Green sand (soft stone?)		4 10	38 0
[Lower	Brown sand		19 0	57 0
Greensand]	Grey sand		16 6	73 6

SELSEY. Park Farm. 1889.

From a section and samples communicated by MESSRS. DUKE & OCKENDEN.

No good water.

			Thickness.	Depth.
			Feet.	*Feet.*
[Drift]	Brickearth		4	4
	Fine beach		8	12
	Sand and beach		7	19
	Lug sand		11	30
	Green [shelly] sand		14	44
[Brackle-sham Beds, 330 feet]	Green [shelly] sand with streaks of light-coloured marl		12	56
	Green [shelly] sand		28	84
	Black clay [sample light-green, smooth, and soapy]. At 90 feet a few inches of substance like coal		6	90
	Sand and clay [laminated (?), with fossils]		12	102
	Green sand		6	108
	Dark clay		16	124
	Green sand [light-grey micaceous clay at 125 feet; green sand at 130 feet; carbonaceous sandy clay at 134 feet]		10	134
	Black clay [light-grey clay, not quite so smooth and soapy as at 84 feet]		3	137
	Black sand [dark-coloured clay and sand]		2	139

		Thickness.	Depth.
		Feet.	*Feet.*
[Brackle-sham Beds, 330 feet]	Sand and clay [laminated clay and sand at 141 feet; green sand at 151 feet]	18	157
	Clay	4	161
	Clay and marl with a little sand [green sand and yellow concretion]	11	172
	Light clay and green sand [green sandy clay at 185 feet; light-grey clay at 212 feet; whitish and pale-yellow clay with darker streaks at 249 feet]	77	249
	Green sand with layers of clay and light-coloured rock [green sand and yellow concretions from 249 to 251 feet]	8	257
	Sand and clay	16	273
	Clay and a little sand	6	279
	Hard clay	3	282
	Green sand and clay	19	301
	Hard black sand-rock	3	304
	[Green] sand and clay	6	310
	Black clay [grey clay at 321 feet]	23	333
	Sand and clay [finer sand at 346 feet]	13	346
	Green sand	9	355
	Sand (layers of) and sand-rock [blacker sand at 357 feet]	2	357
	Sand [with glauconite]	3	360
[London Clay, 192 feet]	Sand and clay [dark-coloured micaceous sandy clay]	16	376
	Black clay [rather sandy at 394 feet]	34	410
	Sand [blacker clay and broken flint]	20	430
	Hard clay [dark sandy clay at 430, 440, 450, 460, and 470 feet; stiffer black clay at 479 feet; more sandy clay at 498 feet; dark-grey or black clay at 500, 502 and 503 feet; black clay with white streaks (crushed septaria?) at 505 feet]	77	507
	Hard white rock (no sample obtained)	$\frac{1}{2}$	$507\frac{1}{2}$
	Black clay	$29\frac{1}{2}$	537
	Bluish sandy clay and brown clay, to running sand	15	552

SHIPLEY. Workhouse. Abandoned (no water).

P. J. MARTIN. "Geological Memoir on a part of Western Sussex," 4to, Lond., 1828, p. 44.

Shaft 75 feet, the rest bored.

Weald Clay. In the middle of the boring a thin shelly bed, and a thicker one at the bottom (two beds of Sussex Marble). 110 feet.

SHORT GATE—SLINFOLD. 81

In Dr. Mantell's "Geology of the South-east of England," 8vo, Lond., 1833, p. 186, a well at Shipley is referred to as having passed through masses of shells at the depth of 30 feet, and also at 100 feet.

Short Gate.

Dr. Mantell. "The Fossils of the South Downs." 4to, Lond., 1822, p. 66. Weald Clay, with two beds of Sussex Marble, 30 feet.

Slaugham. Ashfold House.

Bored and communicated by Messrs. Duke & Ockenden.
Plenty of water, standing at 190 feet from the surface.

	Thickness.	Depth.
	Feet.	Feet.
Old well	—	39
Blue rock	23	62
Coal [lignite]	1	63
Very hard blue rock	10	73
Blue rock and marl	3½	76½
Hard white marl	10½	87
Hard white rock	7½	94½
Very hard blue rock	29½	124
Blue rock and sand	4½	128½
Blue rock and clay	5	133½
Clay and white marl	3½	137
Rock	4	141
Hard rock	3	144
Hard rock and clay	5	149
Hard rock	5	154
Hard rock and a little clay	16	170
Hard rock	35	205
Hard rock and a little clay	4	209
Very hard rock	14½	223½
Very hard sand-rock (spring)	36½	260

Slinfold. Rowfold Farm.

Bored and communicated by Messrs. Duke & Ockenden.

		Thickness.	Depth.
		Feet.	Feet.
	Old Well (the rest bored)	—	59
	Light-coloured clay	6	65
[Weald Clay]	Clay	10	75
	Blue rock	7	82
	Rock and blue marl	46	128

1178. F

SLINFOLD. Rapkins (Mr. W. D. Knight's) E.N.E. of the village.
Made and communicated by MESSRS. A. WILLIAMS & Co.
Shaft 6 feet, the rest bored.
Water-level 63 feet down.   Yield (with a 3-inch pump) 500 gallons an hour.

		Thickness.	Depth.
		Feet.	*Feet.*
[?Horsham Stone]	Brown clay and rock	6	6
	Light-blue clay, with sand-rock	7	13
	Blue clay and shaly rock	2	15
[Weald Clay]	Dark dry hard clay	7	22
	Light-[coloured] dry hard clay	20	42
	Light-blue hard clay	5	47
	Light-brown rock, with clay	6	53
	Dark hard dry clay	6	59
	Light-blue slaty rock	7	66
	Light-[coloured] slaty clayey rock	6	**72**
	Light-brown shaly clay	9	**81**
	Light-brown hard clay	4	85
	Hard light-blue clay	5	90
	Dark-grey shaly clay	6	96
	Light-blue shaly clay	4	100
	Light-blue shaly rock	4	104
	Dark-blue shaly rock	6	110
	Dark-blue shale	6	116
	Light-blue hard clay and rock	7	123
	Blue shaly clay	30	153
	Blue shaly rock	10	163

SOMPTING.

F. Dixon's "Geology of Sussex," Ed. 2, 1878, p. 78.

Gravel, with a little marl - - - - 10 or 12 feet.
Sand, with marine shells of recent species - 6 or 7 „
Chalk, with very good water.

STEYNING. Shelley's Farm, for Mr. Gates.
Made and communicated by MESSRS. DUKE & OCKENDEN.
Water coming in at different places in rocks, stands 28 feet down.

	Thickness.	Depth.
	Feet.	*Feet.*
Hard clay	24	24
Rock	4	28
Hard clay	26	54
Clay, with thin layers of stone	12	66
Hard clay	18	84
Rock	1	85
Hard clay, of varying colours	47	132

STEYNING. Waterworks, *see* UPPER BEEDING.

SUB-WEALDEN BORING, *see* MOUNTFIELD.

TELSCOMBE. Warren Farm. Brighton Industrial School. 1858-1862.

H. WILLETT. In F. Dixon's "Geology of Sussex, Ed. 2, 1878, pp. 115-117. From *Brighton Gazette*, 17th April, 1862.

Shaft 437 feet. Then, at 400 feet, a gallery northward, 7 feet high, 6 wide and 30 long. At 12 feet lower another, to the west, 9 feet high, 6 wide and 30 long. These connected by another, 6 feet high, 3 wide and 20 long. Another heading eastward, 9 feet high, 6 wide and 20 feet long. From these sources only 1,000 gallons of water per day were got.

Then another shaft was made in the eastern gallery, and this continues to the base. [The construction of this well is peculiar. Why the work of sinking a shaft was continued in so inaccessible a place instead of being taken straight down from the surface is hard to understand.]

	Thickness.	Depth.
	Feet.	*Feet.*
[Upper] Chalk, with flints. A thin seam of marl [? 3 feet] at the base - - - - -	418	418
[Middle] Chalk, without flints - - -	212	630
[Lower Chalk] Grey marl, with blue seams -	155	785
[Lower Chalk and partly Upper Greensand?] Blue marl with grey seams - - - -	173	958
Upper Greensand. Firestone without water -	10	968
Gault, 312 feet { Clay, varying from ash-brown to black and bluish-black -	282	1,250
Clay, with seams of green sand, much vegetable matter, wood and pyrites. A sulphurous stench from this -	25	1,275
Brown clay, not effervescing with acid, as the rest of the Gault does, with hard white nodules (? phosphatic) -	5	1,280
[? Lower Greensand or Gault] Greensand with seams of white sand, mixed with pebbles, [? phosphatic nodules] - - - - -	5	1,285
[Lower Greensand] Red sand, touched by a small auger - - - - - - -	-	-

"The beds dip S., and for this reason a deduction of 5 to 10 per cent. should be made from the above figures to get the true thickness. [This would imply a very high dip, of which there is no evidence at the surface.]

On March 16th, 1862, after the workmen had ascended the lower shaft, the thin floor of Gault left at the bottom of the well was broken up, under a pressure of 420 lbs. to the square inch, by the water in the sand below, and the first descending man of the next shift, got into water at 400 feet from the bottom, 32,000 gallons having rushed in during the interval of three-quarters of an hour. The water continued to rise, but it took several days to fill the galleries, and by April 10th it had risen to 945 feet from the bottom [340 from the surface], or 60 feet above low water-mark, when the well held 100,000 gallons."

[As there is an outlier of Reading Beds close to the site, we have here all but the whole thickness of the Chalk. The collective thickness of Middle Chalk, Lower Chalk and Upper Greensand is 550 feet, which seems excessive, even allowing the suggested reduction for dip.]

For Analysis of the water see p. 119.

1178.

THAKEHAM. Merrywood. Mr. Gilbert's.

Bored and communicated by MESSRS. DUKE & OCKENDEN.

Good supply, in the sand, the water rising to within 55 feet of the ground.

		Thickness.	Depth.
		Ft. in.	Ft. in.
Well (old)		—	60 0
[Hythe Beds] {	Hard blue rock	30 0	90 0
	Green sand, to rock	0 10	90 10

THREE BRIDGES STATION, see WORTH.

TICEHURST. Metropolitan Drinking Fountain. Middle of road in centre of village. 1885.

Made and communicated by MESSRS. LE GRAND & SUTCLIFF.

Water-level $12\frac{1}{2}$ feet down. Yield $2\frac{1}{2}$ gallons a minute.

		Thickness.	Depth.
		Feet.	Feet.
Dug (the rest bored)		—	2
	Clay	12	14
	Yellow loamy clay	6	20
	Clay	2	22
	Stone	$3\frac{1}{2}$	$25\frac{1}{2}$
[Weald Clay] {	Blue marl	$4\frac{1}{2}$	30
	Mottled marl	4	34
	Sandstone	$5\frac{1}{2}$	$39\frac{1}{2}$
	Blue marl	1	$40\frac{1}{2}$
	Loamy clay and stone	$2\frac{1}{2}$	43
	Blue marl	23	66

TUNBRIDGE WELLS. L. B. & S. C. Station, see FRANT.

UCKFIELD. Grammar School.

Communicated by MR. SMITH, the Head Master.

Old well 50 feet, the rest bored.

Water rose to the height of 45 feet below the ground. The water in the old well was bad. The supply from the bore-hole is at the rate of 300 to 400 gallons an hour, and the water can be pumped without lowering the head.

		Thickness.	Depth.
		Feet.	Feet.
[Tunbridge Wells Sand] {	Bricked	—	6
	Sandstone	44	50
	Rock (sandstone)	14	64
[Wadhurst Clay]	Blue clay	187	251
[Ashdown Sand. ? in part Wadhurst Clay]	Lighter [-coloured] clay and sand, gradually more sandy	13	264

UCKFIELD. Waterworks, Hempstead Mill.

Made and communicated by Messrs. A. WILLIAMS & Co. With some additions from Mr. H. B. NICHOLS (1890).

Engine-house floor about 100 feet above Ordnance Datum.
Water-level 90 feet above Ordnance Datum. Reduced 20 feet by pumping at the rate of 6,000 gallons an hour, after 3 to 4 hours.

		Thickness.	Depth.
		Feet.	*Feet.*
[Soil and Alluvium]	Yellow clayey alluvial soil	20	20
[Wadhurst Clay, 180 feet]	Blue clay, with thin bands of red clay toward the top	20	40
	Very compact blue clay	20	60
	Very hard blue clay	15	75
	Very hard and very solid blue clay	10	85
	Softer blue clay	5	90
	Hard blue clay, with sandy particles	10	100
	Blue clay	10	110
	Light-grey sandstone	10	120
	Compact blue clay	20	140
	Sandy paste clay	10	150
	Blue rock (? calcareous grit), very hard, with sand in streaks	25	175
	Compact blue clay, very hard	25	200
[Ashdown Sand, 78 feet]	Light-grey sand	30	230
	Light-grey sand, with clay	10	240
	Close-grained sand	15	255
	Fine white sand	5	260
	Hard blue clay	5	265
	White-grey sand, very fine and clear	5	270
	Very pure white sand, with water. Thin bands of red clay toward the top	8	278

UPPER BEEDING. Room Bottom.

Boring (trial) for Steyning water-supply, made and samples communicated by Messrs. DUKE & OCKENDEN.

114 feet above Ordnance Datum.

		Thickness.	Depth.
		Feet.	*Feet.*
Run of the hill	Chalk rubble and flint	15	15
	Grey chalk	25	40
Lower Chalk	Dark-grey marl - at 40 feet Light-grey hard chalk at 55 „ Dark-grey marl - at 65 „ Grey marl - at 69 „ Do. - at 70 „ Dark-grey hard marl at 80 „ Hard grey marls at 85, 88, 97, and 100 feet Dark-grey hard marl at 101 and 102 feet		

WADHURST. Buckhurst Manor Farm (? about two miles westward of the Church).

Made and communicated by MESSRS. LE GRAND & SUTCLIFF.

Water-level 120 feet down.

		Thickness.	Depth.
		Feet.	Feet.
Pit (? the rest bored)		—	5
[? Wadhurst Clay]	Yellow clay	7	12
	Blue clay	11	23
	Yellow claystone	2	25
	Sandstone	4	29
	Yellow clay and stone	38	67
	Hard stone	4	71
	White sandstone	3	74
	Yellow clay and stone	4	78
[? Ashdown Sand]	Brown sandstone	19	97
	Clay and stone	12	109
	Stone	2	111
	Clay and stone	2	113
	Hard stone	2	115
	Clay and stone	3	118
	Hard stone	24	142
	Dark mottled clay	3	145
	Clay and stone	14	159
	Hard sandstone	4	163
	Stone and loamy sand	2	165
	Hard sandstone	20	185
	Loamy clay	2	187
	Hard sandy blue clay	8	195

WALBERTON. Messrs. Ellis & Sons' Brewery.

Section and samples communicated by MESSRS. ELLIS.

		Thickness.	Depth.
		Feet.	Feet.
[Drift]	Loamy sand - about	20	20
[London Clay]	Blue clay - about	30	50
[Reading Beds]	Mottled clay, at about 50 feet [?] Red clay at 128 and 146 feet Lignite and black clay, thin bed between 150 and 160 feet Grey clay, at 170 feet Red clay, at 171, 171½, and 173 feet Hard grey sand, at 174 and 176 feet Mottled sand and clay, at 183 and 185 feet Red clay, at 187, 190, and 192 feet Dark mottled clay, at 193, 194, 196, and 197 feet Hard bed, at 198 feet	148	198
[Upper Chalk]	Mixed chalk and clay, at 199 feet Layers of flint beds and blue clay, from 199 to 214 feet Chalk, at 214 feet Chalk and flints	115	313

The London Clay should probably be thicker, and the Reading Bed thinner, than is given. The first numbered sample was from 128 feet.

WALDRON. In the railway cutting between the north-east end of Heathfield Railway Station and the mouth of the tunnel (60 yards south of Hotel). 1896.

C. DAWSON, *Quart. Journ. Geol. Soc.*, vol. liv., pp. 570, 571.

450 feet above Ordnance Datum.

		Thickness.		Depth.	
		Ft.	in.	Ft.	in.
Dug well	(No record—see next section)	—	—	73	0
	Grey sandy marl-rock	6	0	79	0
	Grey sandy marl-rock, with bands of grey sandstone	17	6	96	6
	Blue sandy marl-rock	4	0	100	6
	Blue sandy marl-rock, with bands of grey sandstone	2	6	103	0
	Blue shale and fossils, grey sandstone with lignite and ironstone	9	0	112	0
	Blue marl-rock, with bands of grey sandstone and ironstone	15	6	127	6
	Blue marl-rock and shale	3	8	131	2
	Blue sandy marl-rock, with occasional ironstone	8	10	140	0
	Hard grey sandstone	1	9	141	9
	Blue sandy marl-rock	1	3	143	0
	Blue sandy marl-rock, with bands of grey sandstone	9	0	152	0
	Blue sandy marl-rock and ironstone	9	6	161	6
	Grey sandstone	1	3	162	9
	Blue sandy marl-rock, with bands of grey sandstone and ironstone	8	6	171	3
[Fairlight Clays]	Grey sandstone (? 9 inches)	1	9	172	0
	Blue sandy marl-rock, with bands of grey sandstone and ironstone	12	0	184	0
	Grey sandstone	0	8	184	8
	Blue sandy marl-rock, with bands of grey sandstone and ironstone	3	4	188	0
	Blue sandy marl-rock, with ironstone	8	3	196	3
	Hard grey calcareous sandstone	1	2	197	5
	Blue sandy marl	0	1	197	6
	Grey calcareous sandstone	0	3	197	9
	Blue sandy marl	0	1	197	10
	Grey calcareous sandstone	0	4	198	2
	Blue marl-rock	0	2	198	4
	Grey calcareous sandstone	0	3	198	7
	Bands of the same, and blue marl-rock	1	11	200	6
	Blue shale, with thin bands of grey calcareous sandstone	7	6	208	0
	Blue marl-rock, with thin bands of blue shale	6	0	214	0
	Blue marl-rock, with thin hard blue shale	1	0	215	0
	Blue shale	2	6	217	6
	Grey sandstone	1	6	219	0
	Blue sandy marl-rock	5	6	224	6

WALDRON—continued.

		Thickness.		Depth.	
		Ft.	in.	Ft.	in.
[Fairlight Clays]	Blue sandy marl-rock, with bands of blue shale	8	0	232	6
	Blue sandy marl-rock and grey sandstone	3	6	236	0
	Blue sandy marl-rock, with nodules of clayey ironstone	8	0	244	0
	Blue sandy marl-rock and shale	13	6	257	6
	Bands of blue sandy marl and shale, with bands of greyish sandstone	7	0	264	6
	Brown and greenish sandy marl, with thin bands of marble	2	6	267	0
	Blue and greenish sandy marl, with bands of shale	3	6	270	6
	Blue and grey sandy marl, with bands of shale	6	0	276	6
	Blue, brown, and greyish marl and shale	10	6	287	0
	Brown and greyish sandy marl and blue shale	13	0	300	0
	Grey sands, marl-rock, and shale (Gas first lighted, 312 feet.)	12	0	312	0
	Blue sandy marl-rock and shale	1	0	313	0
	Greyish limestone	0	1½	313	1½
	Blue sandy marl-rock	2	10½	316	0
	Blue sandy marl-rock, with nodules of grey sandstone	5	0	321	0
	Blue sandy marl-rock, with bands of shale (*Paludina fluviorum*, 333 feet)	18	0	339	0
[Purbeck Beds]	Blue sandy marl-rock, with bands of bituminous shale and broken fossils (*Corbula* and *Cyrena*, 347 feet)	8	0	347	0
	Blue sandy marl-rock, and bands of hard bituminous shale with shells (*Ostrea, Melania, Hydrobia?, Corbula, Cyrena, Cardium*, &c., 353 feet *et seq.*)	6	0	353	0
	Bands of shell-rock and shale	3	6	356	6
	Blue sandy marl, with bands of shale with shells	5	6	362	0
	Blue shale and bands of shell-rock	3	6	365	6
	Shell-rock	0	8	366	2
	Bands of blue shale with shells	1	4	367	6
	Blue shale and hard bands of shells	3	6	371	0
	Blue sandy marl, with bands of shale with shells	6	0	377	0

Little water was found in this boring; but gas has continued to escape, though in March, 1898, the boring was found to be blocked at the depth of 229 feet from the surface.

WALDRON. New Heathfield Hotel. On the southern side of the main road, half-a-mile west of the town. 1893.
C. DAWSON, *Quart. Journ. Geol. Soc.*, vol. liv., p. 569.
493 feet above Ordnance Datum. Sunk 21 feet, rest bored.
Abandoned as unsuccessful.
Water-level 180 feet down. Was higher when the boring was shallower.

		Thickness.	Depth.
		Feet.	*Feet.*
[Ashdown Sand]	Dark brown rusty ferruginous sand, with very thin bands of lignite	5	5
	Light yellow and grey sand, with thin bands of lignite	5	10
	Slate-coloured marl	1	11
	Yellow and white bands of sand	10	21
	Sandstone and blue marl, in layers	11	32
	White sandstone	18	50
	White sandstone and layers of marl or clay	9	59
	Blue sandstone	2	61
	Blue sandstone and layers of marl	10	71
	White and yellow sandstone	5	76
	Blue sandstone and marl	10	86
	Blue marl. First signs of water	3	89
	Sandstone and marl	5	94
[Fairlight Clays]	Blue marl	57	151
	Hard sand-rock	4½	155½
	Blue marl	12½	168
	Hard stone	1½	169½
	Hard blue marl	46½	216
	Hard sand-rock	3	219
	Blue marl (inflammable gas first noticed at 228 feet)	30	249

WANNOCK, see FOLKINGTON & JEVINGTON.

WARNHAM.
"The Geology of the Weald," p. 101.

	Thickness.		Depth.	
	Feet	*in.*	*Feet*	*in.*
Bluish clay	7	0	7	0
Red sandstone	0	9	7	9
Bluish clay	20	0	27	9
Red sandstone	0	8	28	5
Blue clay	15	7	44	0
Hardened blue clay	2	6	46	6
Blue clay	31	6	78	0
Water-bearing bed	1	0	79	0
Blue clay	35	0	114	0
Hard sandy clay	3	0	117	0
Blue clay, with fragments of other formations [?]	27	0	144	0
Red clay, with fragments of red sandstone [?]	8	0	152	0

WARNHAM. Lodge.
Made and communicated by Messrs. Docwra.

Shaft 82½ feet, now filled to 80. Heading to old well at 24 to 29 feet down.

		Thickness.	Depth.
		Feet.	*Feet.*
Soil		1½	1½
[Weald Clay]	Loamy yellow clay	4	5½
	Hard brown rocky marl	13¾	19¼
	Hard blue (marl) clay	6	25¼
	Brown clay	2¾	28
	Blue clay	75½	103½

Water got at the depth of 73 feet was lost at 80 feet. Present supply from a higher level (? 33).

WARNHAM. Kingsfold Estate.
Bored and communicated by Messrs. Duke & Ockenden.

Place Farm, on the hill.
Weald Clay, no water 303 feet.

According to Mr. P. Chasemore there was plenty of surface-water, which, however, was contaminated with salt, making it unfit for use.

At the Crossing gates in the Park, east of the Farmstead.
Weald Clay, no water, 84½ feet.

At the western corner of the Park, by the side of the railway.
Weald Clay, to plenty of bad water, with Epsom salts, 73 feet.

West of the last boring, and under the windmill at the foot of the hill.
Weald Clay - - - - - 30 ⎫
Blue rock, with hard but passable water - 27 ⎬ 57 feet.

According to Mr. P. Chasemore water was found, in the Horsham stone-beds, at the bottom.

WARREN FARM, *see* TELSCOMBE.

WEST DEAN. Trial-boring for the Eastbourne Waterworks. North of the pond.
Made and communicated by Messrs. Isler & Co.
Water-level 5 feet below surface.

		Thickness.	Depth.
		Feet.	*Feet.*
Dug Pit		—	4
[Alluvium]	Clay	8	12
	Running sand	3	15
	Blue clay	27	42
	Rock	1	43
[Upper Chalk]	Chalk and flints	57	100

WEST DEAN. Trial-boring for the Eastbourne Waterworks. In the valley half a mile above the pond.

Made and communicated by MESSRS. ISLER & Co.

Water-level 5 feet below surface.

		Thickness.	Depth.
		Feet.	*Feet.*
Dug Pit		—	6
[Drift]	Mottled clay	5	11
[Drift]	Clay and flints	5	16
[Upper Chalk]	Chalk and flints	84	100

WESTFIELD. Hastings Waterworks. Brede Valley Scheme. Trial-borings and wells.

Communicated by MR. P. H. PALMER, Engineer to the Borough.

1. *Brede Bridge*, just west of the road and south of the stream. About $11\frac{1}{2}$ feet above Ordnance Datum.

The water (found at 83 feet deep) ran over the top of the tube at the rate of 28,000 gallons in 24 hours.

		Thickness.	Depth.
		Feet.	*Feet.*
[Alluvium]	Sandy clay	$3\frac{1}{2}$	$3\frac{1}{2}$
[Alluvium]	Peat	24	$27\frac{1}{2}$
[Ashdown Sand]	Clay, slight traces of sand	15	$42\frac{1}{2}$
[Ashdown Sand]	Clay and sand	8	$50\frac{1}{2}$
[Ashdown Sand]	Blue stone (sandstone)	2	$52\frac{1}{2}$
[Ashdown Sand]	Dense hard mottled clay	7	$59\frac{1}{2}$
[Ashdown Sand]	Clay	2	$61\frac{1}{2}$
[Ashdown Sand]	Pipe-clay	$8\frac{1}{2}$	70
[Ashdown Sand]	Clay and sand	$2\frac{1}{2}$	$72\frac{1}{2}$
[Ashdown Sand]	Very dense clay	3	$75\frac{1}{2}$
[Ashdown Sand]	Clay and sand	2	$77\frac{1}{2}$
[Ashdown Sand]	Clay sand and pebbles	$4\frac{1}{2}$	82
[Ashdown Sand]	Close-grained white sand. Water (increasing to 108 feet)	6	88
[Ashdown Sand]	White sand and clay	6	94
[Ashdown Sand]	White sandstone	14	108
[Ashdown Sand]	Rock marl (firm clayey sand)	8	116
[Ashdown Sand]	Sandstone	8	124
[Ashdown Sand]	Blue marl (firm clayey sand)	4	128
[Ashdown Sand]	Sandstone	6	134
[Ashdown Sand]	Clay	8	142
[Ashdown Sand]	Sandstone	$2\frac{1}{2}$	$144\frac{1}{2}$
[Ashdown Sand]	Clay	$15\frac{1}{2}$	160

WESTFIELD. Hastings Waterworks—*continued*.

2. *Owl's Castle* (N.W. of), 60 feet above Ordnance Datum.

		Thickness.	Depth.
		Feet.	Feet.
	Clay	7	7
	Sand and clay	5½	12½
	Blue marl	11⅔	24
	Brown clay	9	33
	Sandstone	19	52
	Blue marl	44	96
[Ashdown Beds]	Light-blue marl	11	107
	Blue stone	1	108
	Blue marl	6	114
	Blue stone	7	121
	Blue marl	10	131
	Blue stone	1	132
	Blue marl	5¾	137¾

3. Just E. of the footpath, by stream a third of a mile N.N.W. of Crowham. 12 feet above Ordnance Datum.

		Thickness.	Depth.
		Feet.	Feet.
Soil		2	2
	Clay	1	3
[Alluvium]	Peat	3½	6½
	Sandy clay (or loam)	15½	22
	Yellow clay	2	24
	Blue clay marl, with stone at the depth of 54¼ to 55½	46	70
	Light-brown silt (or sand)	1	71
	Blue stone	1	72
	Sand-rock	9⅓	81⅓
Ashdown Sand]	Blue marl (or clay) and sand-rock (rather sandy clay) inter-mixed	18⅔	100
	Hard white sand-rock and clay	23	123
	Hard brown (sandy) clay (or sand-rock and clay)	18	141
	Greyish clay	20	161

[Can there be Wadhurst Clay here? Nothing but Ashdown Sand is shown on the Geological Survey Map.]

WESTFIELD. Hastings Waterworks—*continued*.

4. *Redley Farm*. Well. By the side-stream, less than a sixth of a mile south-westward of the house.

Over 29½ feet above Ordnance Datum. Original surface 3 feet lower.

		Thickness.		Depth.	
		Ft.	*in.*	*Ft.*	*in.*
[Alluvium]	Soil, loam and peat	4	6	4	6
	Silt and clay	4	6	9	0
	Sand-rock and ironstone	2	3	11	3
	Blue marl or clay. Two inches of beach-stones at the depth of 38½ feet. Below the pebbles the beds dip 60° to S.S.W.	89	11	101	2
	Blue stone	1	10	103	0
	Blue marl or clay (*Unio antiquus* at the depth of 112 feet)	13	0	116	0
	Blue stone	0	6	116	6
	Sandstone and water	5	6	122	0
	Blue clay	5	0	127	0

[Can there be Wadhurst Clay here?]

5. In the marsh five-twelfths of a mile above Brede Bridge, 10¼ feet above Ordnance Datum.

A shallow sump (? about 15 feet), the rest bored [descriptions of specimens in these brackets].

		Thickness.		Depth.	
		Ft.	*in.*	*Ft.*	*in.*
[Alluvium]	Soil and peat	12	0	12	0
	Very soft, muddy (?sandy) clay [fine yellowish-buff clayey sand]	23	0	35	0
	Soft brown sand and clay	4	0	39	0
	Soft ironstone (or sandstone) and shale [buff loam, and pieces of iron sandstone], (dirty sand, and gravel of Wealden sandstone)	2	6	41	6
	Brown sandy loam [fine buff clayey sand or sandy clay]	3	0	44	6
	Blue clay and pebbles	0	6	45	0
	Blue clay [pale grey clay, at 66 feet, not calcareous]	85	0	130	0
	Blue stone (hard grey calcareous grit or sandstone)	2	6	132	6
	Running [fine grey] sand and water	0	8	133	2
	Blue clay [grey, slightly calcareous]	30	7	163	9

WESTFIELD. Hastings Waterworks.—*continued*.

		Thickness.		Depth.	
		Ft.	in.	Ft.	in.
	Hard bluestone (calcareous grit)	8	9	172	6
	Blue marl [grey calcareous clay and pale soft sandstone]	18	0	190	6
	Blue stone [grey, calcareous]	2	6	193	0
	Blue marl [grey calcareous clay, and then somewhat sandy, but not calcareous]	15	0	208	0
	Sand-rock [fine grey sand at 213 and 213½ feet, fine pale buff clayey sand at 238, brownish clay at 243½, pale grey clay at 247½, pale buff and grey clay at 249½. Not calcareous]	57	0	265	0
	Brown clay shale [buff clay, slightly sandy, not calcareous]	24	6	289	6
	Sand-rock (or stone) [fine buff sand, partly compacted]	29	0	318	6
	Clay [pale brownish-grey, not calcareous]	3	0	321	6
	Sand and clay [pale brownish-grey sand]	1	0	322	6
	Clay [pale brownish-grey, not calcareous]	1	6	324	0
	Stone sand [pale brownish-grey] and clay	4	9	328	9
	Clay [very pale (whitish), not calcareous]	1	3	330	0
	Stone, sand and clay	1	6	331	6
	Clay	2	0	333	6
	Stone, sand and clay [fine buff sand]	0	6	334	0
	Clay [very pale clay, not calcareous]	—		334	6

Water met with at the depth of 132 feet 8 inches, and it overflowed at the rate of 50,000 gallons in 24 hours. A pump of 6 inches diameter placed in the borehole, and worked for 24 hours, could not lower the water more than to 14½ feet below the ground, the yield being still 50,000 gallons.

A well 52 feet deep has been made close by.

6. In the marsh, about a third of a mile N. of E. from **Rock's** Farm.

About 14¾ feet above Ordnance Datum.

A shallow sump (? about 15 feet), the rest bored.

Pumping here affected Nos. 5 and 4.

Water from the beds of sand-rock (and loose sand) between **62** and **119** feet down overflowed at the rate of 70,000 gallons (in 24 hours?). It

WESTFIELD. Hastings Waterworks.—*continued*.

has ceased to overflow, and has gradually lowered [? through pumping from the well close by].

		Thickness.		Depth.	
		Ft.	in.	Ft.	in.
[Alluvium]	Soil, peat, &c.	16	0	16	0
	Peat	6	0	22	0
	Clay	5	0	27	0
[Ashdown Sand]	Yellow sandstone (broken, gravelly)	2	6	29	6
	Grey sandstone (? with fish scales)	4	3	33	9
	Blue stone (calcareous grit). Water ran over the top	1	7	35	4
	Blue marl [? 19 ft. 8 in.]	20	5	55	9
	Running sand (water) [?4 feet]	3	3	59	0
	Clay	3	0	62	0
	Yellow sand-rock	2	6	64	6
	Sand-rock, with 9 inches of clay at the base, probably a fissure	16	0	80	6
	Hard grey sand-rock	19	6	100	0
	Sand-rock	19	0	119	0
	Brown clay (shale) [? 24 ft.]	27	0	146	0
	Sharp brown sand, or sand-rock [? 22 ft.]	19	0	165	0
	Sand and clay	3	0	168	0
	Sand-rock	31	6	199	6
	Clay	3	0	202	6
	Sand and clay, chiefly the former	7	6	210	0
	Sand-rock	10	0	220	0

In another account the top two beds are given as Clay 4 feet, Peat 18; and the beds below 168 feet as Sand-rock 17½, Clay 14½, Sand and clay 2½, Sandstone 3½, this account stopping at the depth of 206 feet.

On touching the sand-rock below 146 [or 143] feet, water burst up in the well close by from a crack or fissure in the bottom, then 54 feet down, and the yield (of the well) increased from 432,000 to 576,000 [gallons in 24 hours].

7. Well, about 30 feet from No. 6. Has yielded a large amount of water.

		Thickness.	Depth.
		Feet.	Feet.
Soil		1	1
Alluvium	Clay	4	5
	Peat	6	11
[Ashdown Sand]	Silt and loose stones	13	24
	Sand and blue stone, in small pieces, mixed	6	30
	Blue marl	5½	35½
	Thin shaly rock	1	36½
	Silt and loose marl, mixed	11½	48
	Sand-rock. Large vents filled with running sand	6	54

WESTFIELD. Hastings Waterworks—*continued*.

An account of this boring from MESSRS. ISLER differs in the following particulars, giving some fuller detail in parts.

Water started overflowing at 56 ft. down. During work between the depths of 56 and 101 feet the springs blew out the bottom of the well, 30 ft. from the boring. Yield increased 150,000 gallons a day more at 147 feet, and the amount at no time reached from 35,000 to 40,000 an hour.

	Thickness.		Depth.	
	Ft.	*in.*	*Ft.*	*in.*
	—	—	29	6
Light [coloured] sandstone	1	4	30	10
Very hard sandstone	5	9	36	7
Blue stone	5	0	41	7
Marl and stone	1	6	43	1
Hard yellow sandstone	3	0	46	1
Hard beds of sand and blue stone	1	11	48	0
Hard sandstone	1	0	49	0
Shale and marl	2	9	51	9
Sand	5	9	57	6
Sandy clay	2	6	60	0
Undescribed	—	—	64	6
Hard sandstone	19	6	84	0
Light [coloured] clay and stone	5	4	89	4
Sand-rock	28	8	118	0
Black shale	22	0	140	0
Layers of hard rocks	0	6	140	6
Black shale	5	6	146	0
Sand-rock	74	0	220	0

8 (called 4 by MR. ELWORTHY *in* "The Hastings Water Supply Past and Present." 8vo. 1894.) By the River Brede, north of Forge Bridge. 1896.

Water, from the sandstone, overflowed and flooded the field.

	Thickness.	Depth.
	Feet.	*Feet.*
Alluvium and clay	42	42
Iron-sandstone	13	55
Clay	8	63
Grey sandstone (like that from which the water is got in Nos. 5 and 6), with here and there a thin layer of clay	153	216
Light-brown shale	13½	229½

WEST FIRLE.   Bushy Lodge.
Made and communicated by MESSRS. DUKE & OCKENDEN
Water was finally obtained from the Lower Greensand.

		Thickness.	Depth
		Feet.	Feet.
[Gault]	Black clay with shell fragments	327	327
[Lower Greensand]	Green sandy clay (as at 170 feet on Ringmer Green)	12	339
[Weald Clay]	Light-grey shaly clay	90	429

WESTHAM. Langley Farm (N.E. of Eastbourne). Two Borings.
Made and communicated by MESSRS. LE GRAND & SUTCLIFF.
No supply found in either.

1.

		Thickness.	Depth
		Feet.	Feet.
Soil		2	2
[Weald Clay]	Loam	8½	10½
	Sandy loam	8½	19
	Weald clay	10	29
	Weald clay and green sand	25½	54½
	Light-grey shaly clay	68½	123
	Light-blue clay and stone	3	126

2. A short distance further from the sea.

		Thickness.	Depth.
		Feet.	Feet.
Soil		1½	1½
[Weald Clay]	Grey sand	4	5½
	Red sand	1½	7
	Brown sand	2	9
	Coloured [mottled] loamy sand	3	12
	Brown live sand	6	18
	Brown loamy sand	9½	27½
	Black sandy clay	13½	41
	Black clay and green sand	34	75
	Weald clay and green sand	23	98
	Weald clay	43	141
	Brown sand	6½	147½
	Weald clay	18	165½
	Light blue clay and stone	9½	175
	Dark brown clay	5	180
	Coloured [mottled] clays	16	196
	Greenish clay with white streaks	5	201
	Brown and blue shaly clay	11	212

WESTHAM. Stone Cross. Mr. Marsden's. 1876.
Made and communicated by MESSRS. S. F. BAKER & SONS.
Good supply, from the sand-rock.

		Thickness.	Depth.
		Feet.	Feet.
[Weald Clay]	Old well [? all clay]	—	65
	Hard dark clay	125½	190½
[Tunbridge Wells Sand]	Hard sand-rock	6	196½

WILLINGDON. Park Croft, near Willingdon Mill. For new Infectious Hospital and Workhouse.

Communicated by MESSRS. DUKE & OCKENDEN.

Old well 60 feet. Then bored.
At 75 feet a little water, shut out by tubes.
Dark greenish clay with very little sand at 150 feet.

WINCHELSEA. Marsh W. of the town.

Shaft, 18 feet diameter.
Yield, 3,000 to 4,000 gallons a day.

		Thickness.	Depth.
		Feet.	Feet.
Soil		2	2
	Unrecorded	3	5
	Loose sand	6	11
[Ashdown Sand]	Rather compact rock	4	15
	Blue marl	2½	17½
	Soft sandstone	6	23½

A letter from Mr. W. Martindale (1890), apparently referring to this well, states that the water is opalescent, and contains a flocculent sediment. The old town-well is 100 feet deep, but has not much water.

WITHYHAM. See p. 102.

WIVELSFIELD. Tawning's Place. Mr. Denman's. 1896?
Made and communicated by MESSRS. DUKE & OCKENDEN.
Shaft 82 feet, the rest bored. Abandoned. No water

	Thickness.	Depth.
	Feet.	Feet.
Soft sand-rock	82	82
Hard sandstone rock	3	85
Stiff slaty clay	8½	93½
Hard rock	6½	100
Slaty rock	32	132
Hard clay	12½	144½
Hard blue rock	6	150½
Hard chalky rock	2½	153
Very hard rock	3	156
Very hard dry clay	2	158

WORTH. Three Bridges Station, western side. About 1887.
Note by Mr. Topley.
Well, through sandstone, 35 feet.
Water rises to 14 feet from the surface. Could not be lowered below 4 feet from the bottom.
Good water. 120,000 gallons a day got.

WORTH. Copthorne. For Mr. Whitchurch.
Communicated by Messrs. G. Isler & Co.
Supply abundant.

		Thickness.	Depth.
		Ft. In.	Ft. In.
Shaft (the rest bored)		—	49 0
[Tunbridge Wells Sand]	Rock	8 5	57 5
	Clay	2 0	59 5
	Rock	42 1	101 6

WORTHING. Chippendale House, Chenwood Road.
Made and communicated by Messrs. Duke & Ockenden.
Water stands 21 feet down, but does not come in very freely.

		Thickness.	Depth.
		Feet.	Feet.
[? Drift, London Clay, and Reading Beds]	Reading bed	76	76
	Grey sand	18	94
	Sand-rock and flints	41	135
[Upper Chalk]	Flint and chalk	40	175
	Hard rock	2	177
	Flint and chalk	8	185
	Flint and sand-rock	7	192
	Chalk and flints	53	245

WORTHING. Mr. Cornden's New Greenhouse, a few hundred yards north of the railway and close to the lane from Chenwood Road to Broadwater. Boring. 1896.
Made and communicated by Messrs. Duke & Ockenden.
Water found at the bottom, stands 7 feet down.

		Thickness.	Depth.
		Feet.	Feet.
[Reading Beds]	Sand and clay	8	8
	Blue and red clay	23	31
	Hard clay	51	82
	Hard clay and gravel [? flints]	5	87
	Hard chalk	17	104
[Upper Chalk]	Hard chalk and flint	19	123
	Hard chalk and hard rock	12	135
	Chalk and flints	107	242
	Hard chalk	48	290

WORTHING. Mr. Page's New Greenhouses, in a field east of the above, opposite private level crossing. Boring. 1896?

Made and communicated by MESSRS. DUKE & OCKENDEN.

Water started coming in at 170 feet, stands 5 feet down.

		Thickness.	Depth.
		Feet.	*Feet.*
[Reading Beds]	Hard red and mixed clay	102	102
[Upper Chalk]	Hard chalk	23	125
	Hard chalk and flints	85	210

WORTHING. Waterworks.

Shaft 60 feet (communicated by MR. BLAKER), the rest bored (communicated by the company).

		Thickness.	Depth.
		Feet.	*Feet.*
[Drift]	Soil	—	—
	Clay	5	5
	Shrave	5	10
	Marl	10	20
	Rubble chalk	50	70
	Solid chalk	5	75
[Upper Chalk]	Chalk, with 1 foot of flints at top and at base	6	81
	Very hard chalk	9	90
	Chalk with flints, 6 inch layer at top	6	96
	Chalk, with 1 foot of flints, and a fissure with water at the base	10	106
	Chalk with flints	64	170
	Chalk, with flints at 176-7, 184-5, 188-9, 195-6, 204-4½, 215-6, 225-6, 235-6, 245-6, 264-5, 283-4 (grey), 300-1, 303-4 (with much water), 306-7, 309—10, 312-2½, 314-5, 318-9, 321-2, 324-5, 328-8½, 330-1, 334-4½, 337-8, 340-1, 343-4, 347-7½, 349-50, 353-4, 356-7, 360-1, 363-4, 365-5½, 367-8, 371-2, 375-6, 379-80, 382-3, 385-5½, 387-8, 390-1, 393-4, 397-8	230	400

For Analysis of the water see p. 120.

WORTHING.

WORTHING. West Worthing Waterworks. 1887.

Made and communicated by MESSRS. LE GRAND & SUTCLIFF.

Water-level 8 feet down (May).

Top ground	3	} 100 feet.
Chalk and flints	97	

WORTHING. West Worthing. 8 chains N.N.W. of the Station.

Bored, and samples communicated by MESSRS. DUKE & OCKENDEN.

		Thickness.	Depth.
		Feet.	Feet.
[Drift]	Sand and shingle - about	34	34
[Reading Beds]	Red clay at 51; green clay at 53; red clay at 54 and 58; mottled clay at 103 and 104; grey clay at 126; lignite at 128; mottled clay at 133, 135, and 137; white and red clay at 138	102	136
[Upper Chalk]	Chalk and black flint	39	175

POSTSCRIPT.

BEXHILL. New Well for Waterworks, 400 feet east of old Well. 1891.

Communicated by Mr. W. B. LEWIS.

114 feet above Ordnance Datum.

Well 8 feet diameter to 196 feet; heading driven to east 180 feet, and then borehole put down.

Supply 48,000 gallons a day.

	Thickness.	Depth.
	Feet.	Feet.
Hard sandy clay	47	47
Sand	2½	49½
Blue marl	50	99½
Sand in beds of 3 or 4 ft. thick. Some water	15	114½
Blue marl	106½	221
Sandstone. Water at 21 ft.	64	285
Brown clay	15	300
Sandstone	27½	327½
Light-blue clay	1	328½
Sandstone	5¾	334¼
Light-blue clay	3¼	337½
Sandstone	2½	340¼
Sandy clay	9¾	350

WITHYHAM. Crowborough Warren. Two **borings**.
Communicated by Mr. C. DAWSON.

1. Nearly **305 feet** above Ordnance Datum. A few **yards** East of the South-Eastern corner of **the** Mill-pond.

Large supply of water tapped 94 feet down and rose to within 23 feet of the ground. Tested up to 50 gallons a minute (at least) after three days and nights pumping. Water rose to within 18½ feet of the surface.

		Thickness.	Depth.
	Yellow clay, with boulders of sandstone	*Feet.* 6	*Feet.* 6
	Blue and red mottled clay	13	19
	Hard sand-rock	5	24
	Sand-rock	11	35
	Blue marl	4	39
	Sand-rock (water)	8	47
	Yellow clay	2	49
	Sand-rock	4	53
	Blue marl	1	54
[Ashdown Sand.]	Grey sand-rock. Gypsum-bands	2	56
	Grey sand-rock	9	65
	Blue marl	6	71
	Sand-rock	5	76
	Blue marl	2	78
	Sand-rock	6	84
	Blue marl	1	85
	Sand-rock	1	86
	Blue marl	5	91
	Drab sand-rock	20	111
	Yellow clay	½	111½

2. Nearly 277 feet above Ordnance Datum. 400 feet N. of the benchmark (295 feet) at the Northern end of the Mill-pond.

At the depth of 67½ feet (?) water stood 11 feet down. On reaching the **bottom** the water-level sank a foot and remained so on testing.

		Thickness.	Depth.
	Marly clay, with grey sandstone-boulders	*Feet.* 10	*Feet.* 10
	Marly clay, with hard sandstone boulders veined with iron	5	15
	Hard sand-rock	4	19
	Gravelly clay, friable **and** pervious	5	24
	Grey and red sandy **rock**	7½	31½
[? All Ashdown Sand, or part Fairlight Clay.]	Blue marly clay	22	53½
	Blue clay	4	57½
	Sandstone with water	4	61½
	Blue marl	2	63½
	Sandstone with water	4	67½
	Blue marl	4	71½
	Sandstone with layers of marl	10	81½
	Red sand-rock. Much water	2	83½
	Blue-marl	1	84½
	Sand-rock	2½	87

ANALYSES OF WATERS.

ANGMERING. Decoy. (1.) From a Well, February 1895. (2.) From a Borehole, March 1895.

Analyses by R. A. CRIPPS. Communicated by MESSRS. DUKE & OCKENDEN.

	1.		2.	
	Grains per Gallon.	[=Parts per 100,000.]	Grains per Gallon.	[=Parts per 100,000.]
Total Solids	67·	[95·7]	26·5	[37·9]
Chlorine	9·9	[14·1]	1·9	2·7
Ammonia	·0112	·016	·0021	·003
Albuminoid of Ammonia	·0098	·014	·0007	·001
Nitrites absent. Nitrates	Excessive quantity.		Trace.	

(1.) Hardness 31·6° [45·14]. This **water is unsafe** for drinking purposes, the quantities of chlorides and nitrates are excessive, and the saline and organic ammonia are both too great. All these indicate organic pollution. Microscopic examination of the sediment yielded equally unsatisfactory results, organic débris both animal and vegetable.

(2.) **This water is of** excellent quality, rather hard, but remarkably free from **organic matter**. Temporary hardness 13·7° [19·57], permanent 3·1° [4·43]. Total 16·8° [24·0].

ARUNDEL. Park. (1) Spring feeding the Swanbourne [draining the uncultivated land of Arundel Park], October 13, 1873. (2) Spring near the Lodge, October 13, 1873.

Rivers Pollution Commission, 6th Report, 1874, p. 122.

Water from the [Upper] Chalk. Temperature 11·3° C. Clear and palatable. [This water is now supplied to the town of Arundel.]

	Parts per 100,000.	
	(1.)	(2.)
Total Solid Impurity	26·30	26·28
Organic Carbon	·054	·037
Organic Nitrogen	·009	·007
Ammonia	—	—
Nitrogen as Nitrates or Nitrites	—	·080
Total Combined Nitrogen	·009	·087
Chlorine	2·10	2·20
Hardness Temporary	18·1	16·1
,, Permanent	4·3	8·1
,, Total	22·4	24·2

The Commissioners add (p. 124), "We have only met with one sample of spring water from the chalk (the spring feeding the *Swanbourne* in Arundel Park, Sussex) which exhibits no trace of evidence of previous pollution with organic matter of animal origin; but in no single case was this evidence sufficiently **strong to** place the sample in the category of **suspicious waters.**"

ARUNDEL. From a Borehole at Offham Farm.

Analysis by R. A. CRIPPS. Communicated by MESSRS. DUKE & OCKENDEN.

	Grains per Gallon.	[= Parts per 100,000.]
Total Solids	44·8	[64·]
Chlorine	10.1	[14·4]
Ammonia	·08 ⎫ parts per	—
Albuminoid Ammonia	·05 ⎭ million.	—
Organic Matter	3·2	[4·57]
Hardness	25°	[35·5]

"Moderately large amount of Chlorides. Organically pure for drinking."

BALCOMBE. Mid Sussex Water Company. October, 1897. (1) Standpipe from Filter Bed. (2) Main near Balcombe Station.

Analyses by R. H. HARLAND.

[Water from the Tunbridge Wells Sand.]

	Grains per Gallon.	[=Parts per 100,000.]	Grains per Gallon.	[=Parts per 100,000.]
Suspended Matter	Very slight trace.		None.	
Temporary Hardness	5·	[7·14]	5·	[7·14]
Permanent Hardness	7·	[10·]	7·	[10·]
Total Hardness	12·	[17·14]	12·	[17·14]
Total Solid Matter	28·0	[40·]	25·2	[36·]
Loss on Ignition	2·8	[4·]	2·1	[3·]
Total Mineral Matter	25·2	[36·]	23·1	[33·]
Chlorine equal to Sodium Chloride	3·2	[4·57]	3·1	[4·43]
Lead, Copper, Iron (in solution)	None		None	
Phosphoric Acid	None		None	
Nitrogen as Ammonia	·0042	[·006]	·0027	[·0038]
Nitrogen as Albuminoid Ammonia	·0013	[·0018]	·0018	[·0026]
Nitrogen as Nitrates	·0560	[·08]	·0672	[·096]
Oxygen absorbed by Organic Matter:—				
In 5 Minutes	Nil.		Nil	
In 4 Hours	·0168	[·024]	·0140	[·02]

"The water as pumped from the well contains a small trace of iron in solution, which rapidly deposits out an exposure to air and light. It is subjected to sand filtration (which entirely removes the trace of iron)."

BEDDINGHAM. Courthouse Farm. March, 1897.

Analysis by Dr. J. A. VOELCKER. Communicated by Mr. T. W. PICKARD.

[Water from the Lower Greensand.]

	Grains per Gallon.	[=Parts per 100,000.]
Total Solid Residue	31·08	44·7
Oxygen absorbed	·060	·086
Lime	1·12	1·6
Magnesia	·30	·43
Sulphuric Acid	1·82	2·6
Nitric Acid	trace	
Chlorine	2·01	2·87
= Chloride of Sodium	3·32	4·74
Free Ammonia	·018	·0257
Albuminoid Ammonia	·006	·0087

"This was yellow coloured but was practically free from deposit. It is a very different water to 1 and 2 [Newhaven Waterworks and Glynde Butter Factory] for, while it has more solid matter than either of the others, there is very little lime or magnesia, and less chlorides and nitrates. Alkaline carbonates appear to be the principal solid constituents composing the residue. The water, accordingly, is one of a soft nature. It contains more dissolved organic matter, which gives rise probably, in measure, to the high amount of Ammonia shown. This latter, however, being unaccompanied by any excess of chlorides, and nitrates being entirely absent, I do not attribute to any objectionable polluting matter, and the water, though one of a peculiar nature, may, I think, be safely used as a drinking supply."

BRIGHTON. Waterworks. Goldstone Bottom Well. Jan. 18th, 1873.

Rivers Pollution Commission, 6th Report, 1874, p. 99.

Water from headings in the [Upper] Chalk at 160 feet.

Temperature 9·6° C. Clear and palatable.

	Parts per 100,000.
Total Solid Impurity	30·24
Organic Carbon	·048
Organic Nitrogen	·009
Ammonia	—
Nitrogen as Nitrates and Nitrites	·644
Total Combined Nitrogen	·653
Chlorine	3·10
Hardness Temporary	14·8
" Permanent	6·4
" Total	21·2

BRIGHTON. Waterworks. **Lewes Road Well.** Jan. 18th, 1873.
Rivers Pollution Commission, 6th Report, 1874, p. 99.
Water from headings in the Chalk at 100 feet.
Temperature 10·0° C. Clear and palatable.

	Parts per 100,000.
Total Solid Impurity	32·40
Organic Carbon	·055
Organic Nitrogen	·011
Ammonia	—
Nitrogen as Nitrates **or Nitrites**	·989
Total Combined Nitrogen	1·000
Chlorine	3·70
Hardness Temporary	14·6
,, Permanent	6·9
,, Total	21·5

BROADWATER. From a Borehole in the Chalk.
Analysis by R. A. CRIPPS. Communicated by MESSRS. **DUKE** & OCKENDEN.

	Grains per Gallon.	[=Parts per 100,000.]
Carbonate of Lime	15·4	22·
Sulphate [of Lime ?]	1·9	2·714
Chlorine	3·1	4·43
Organic Matter	2·4	3·43
Free Ammonia		0·1 parts per
Albuminoid Ammonia		0·2 million.

Hardness, total 16° [22·85], permanent 8·4° [12·0].

"A good sample of a Chalk supply. Chlorine moderate, for the locality A drinking water of perfect purity."

BURPHAM. From a Well.
Analysis by R. A. CRIPPS. Communicated by MESSRS. **DUKE & OCKENDEN.**

	Grains per Gallon.	[=Parts per 100,000.]
Total Solids	26·	37·
Chlorine	2·1	3·
Ammonia	·0007	·001
Albuminoid Ammonia	·0028	·004
Nitrites	absent	
Nitrates	traces	
Hardness Temporary	11·8°	16·85
,, Permanent	4·5°	6·4
,, Total	16·3°	23·25

"Water of good quality, free from organic pollution and contains only a moderate amount of dissolved saline substances. Microscopic examination satisfactory."

ANALYSES.

CHICHESTER. From a Well.

Analysis by R. A. CRIPPS. Communicated by MESSRS. DUKE & OCKENDEN.

[Water apparently from Upper Chalk].

	Grains per Gallon.	[=Parts per 100,000.]
Total Solids	20·5	29·3
Chlorine	1·4	2·
Ammonia	·00056	·0008
Albuminoid Ammonia	·0007	·001
Nitrites	absent	
Nitrates	trace	
Hardness Temporary	13·8°	19·7
,, Permanent	3·2	4·6
,, Total	17·	24·3

"A first-class **water for drinking purposes, free from organic pollution** and containing only a very **moderate quantity of dissolved mineral** matter. Microscopic examination **very satisfactory.**"

CRAWLEY. Trial-boring for **Waterworks**.

Analysis by DR. T. STEVENSON, 1898.

Communicated by the **Waterworks** Company.

[Water from the **Tunbridge Wells** Sand].

	Grains per Gallon.	[=Parts per 100,000.]
Total Solid Matter	31·64	45·2
Loss on ignition	·56	·8
Combined Chloride (=**Common Salt** 1·62)	·98	1·4
Nitrogen as Nitrates (no **Nitrites**)	·05	·07
Carbonate of Sodium	25·22	36·03
Ammonia	·02	·03
Albuminoid or Organic **Ammonia**	·0025	·0036
Oxygen required to **oxidise the organic matter**	·057	·081
Hardness	·5°	·7

"**The water** was free from odour and when viewed in bulk of **a yellow colour and turbid.**" It "**is** well fitted for domestic **use.** It is very soft **and free from** organic contamination." As with "all waters from fresh borings the ammonia is rather high, but this is immaterial." The water is exceptional in containing so much carbonate of sodium, but "in this respect it resembles the **waters from** some Mid-Kent wells. I have not **found** the presence **of** this **quantity of** carbonate of sodium of any detriment except that such waters **act freely on** ordinary compo-metal taps."

EASTBOURNE.   Holywell Springs.   September 18, 1895.
Analysis by DR. THOS. STEVENSON.
[Water from Middle and Lower Chalk.]

	Grains per Gallon.	[= Parts per 100,000.]
Total Solid Matters	20·16	[28.8]
Loss on Ignition	1·40	[2·0]
Combined Chlorine	2·38	[3·4]
Equal to Common Salt	3·92	[5.6]
Nitrogen as Nitrates	·23	[·33]
Nitrites	None.	
Heavy Metals (Lead, Copper, Zinc, &c.)	None.	
Ammonia	None.	
Albuminoid or Organic Ammonia	·001	[·0014]
Oxygen required to oxidise the Organic Matter	·006	[·0086]
Hardness Temporary	10·5°	[15·0]
,, Permanent	2·5°	[3·57]
,, Total	13·0°	[18·57]

"The results of chemical analysis are quite satisfactory, since there is no evidence of pollution with sewage, or contamination with injurious metals. The water is of high organic purity, and of moderate hardness most of which is due to chalky matters."

EASTBOURNE.   Spring above the town.   February 22, 1873.
Rivers Pollution Commission, 6th Report, 1874, p. 123.

[Water apparently from the Lower Chalk or base of the Middle Chalk.] Temperature 10·3° C.   Clear and palatable.

	Parts per 100,000.
Total Solid Impurity	36·46
Organic Carbon	·070
Organic Nitrogen	·011
Ammonia	·001
Nitrogen as Nitrates or Nitrites	·736
Total Combined Nitrogen	·748
Chlorine	3·90
Hardness Temporary	24·2
,, Permanent	7·7
,, Total	31·9

EASTBOURNE.   Well at Waterworks.   [Old Well N. of Engine House.] Rivers Pollution Commission, 6th Report, 1874, p. 97.

Water from [the Upper Greensand at] 100 feet.   Temperature 10·0° C. Slightly turbid.   Palatable.

	Parts per 100,000.
Total Solid Impurity	43·12
Organic Carbon	·058
Organic Nitrogen	·010
Ammonia	·004
Nitrogen as Nitrates and Nitrites	·130
Total Combined Nitrogen	·143
Chlorine	10·00
Hardness Temporary	13·8
,, Permanent	7·1
,, Total	20·9

EASTBOURNE. Star Brewery Company's Well. October 14th, 1895. Analysis by DR. A. WYNTER BLYTH.

	Grains per Gallon.	[= Parts per 100,000.]
Chlorine	3·30	4·71
Free Ammonia	·0003	·0004
Albuminoid Ammonia	·0022	·0031
Nitrogen as Nitrates	1·19	1·7
Oxygen consumed in 15 minutes	·0497	·071
Oxygen in hour at 100° C.	·1515	·2164
Alkalinity expressed as Ca Co 3	15·30	21·86
Hardness (in Degrees)	22·0	31·4
Hardness (after Boiling)	13·0	18·57
Total Solids	25·2	36·0
Loss on Ignition	9·10	13·0
Metals	Absent	
Sulphates	More than traces.	
Organic Carbon	4·6 } parts per	
Organic Nitrogen	0·14 } million.	

"The appearance of this water when viewed through a two-foot tube was that of a clear liquid. No deposit fell on standing and the microscopical appearance was negative. The sample has all the characters of a deep chalk spring, and, considered as such, it must be returned as a fairly pure water."

EASTBOURNE. Waterworks. March 20, 1897.

(1) Holywell. (2) Friston Well. (3) Wannock Well.

Analysis by SIR E. FRANKLAND.

[(1) is from base of Middle Chalk. (2) is from Upper Chalk. (3) is from shattered Lower Chalk.]

	Parts per 100,000.		
	(1) Holywell.	(2) Friston.	(3) Wannock.
Total Solid Matters	33·24	37·08	25·24
Organic Carbon	·055	·041	·043
Organic Nitrogen	·010	·012	·011
Ammonia	—	—	—
Nitrogen as Nitrates and Nitrites	·664	·656	·096
Total Combined Nitrogen	·674	·668	·107
Chlorine	5·2	4·3	2·6
Hardness Temporary	14·5	16·9	15·2
,, Permanent	7·3	7·0	4·2
,, Total	21·8	23·9	19·4

"For Chalk waters they are all of moderate hardness, the Wannock Well remarkably so."

EAST GRINSTEAD. From a Well at Brook House.

Analysis by R. A. CRIPPS. Communicated by MESSRS. DUKE & OCKENDEN.

	Grains per Gallon.	[= Parts per 100,000.]
Total Solids	19·2	27·4
Chlorine	1·9	2·71
Ammonia	·066	·094
Albuminoid Ammonia	·034	·048
Organic Matter	3·7	5·28
Hardness	11°	15·5

"The hardness, chlorides, and organic matter are from the clear water, after subsidence. Water fit for domestic use and for drinking, after subsidence or filtration."

GLYNDE. Butter Factory.
Analysis by PROF. A. DUPRÉ. 1891.

Water clear, almost colourless, inodorous, with no deposit.

	Grains per Gallon.	[= Parts in 100,000.]
Oxygen absorbed from Permanganate	·023	·033
Total dry Residue (white)	28·28	40·40
Chlorine	2·17	3·10
Nitric Acid (no Nitrous or Phosphoric)	2·	2·86
Ammonia	—	—
Albuminoid Ammonia	—	—
Poisonous Metals, minute trace		

The residue blackens on ignition, scarcely perceptible.
Hardness before boiling 19° [27·14].
Hardness after boiling 4° [5·71].
This water is of exceptional purity

GLYNDE. Butter Factory. March, 1897.

Analysis by DR. J. A. VOELCKER. Communicated by MR. T. W. PICKARD

[Apparently a mixture of waters from the Lower Chalk and the Upper Greensand.]

	Grains per Gallon.	[= Parts per 100,000.]
Total Solid Residue	29·96	42·8
Oxygen absorbed	·053	·076
Lime	11·48	16·4
Magnesia	·60	·86
Sulphuric Acid	1·73	2·47
Nitric Acid	2·38	3·4
Chlorine	2·85	4·07
= Chloride of Sodium	4·70	6·91
Free Ammonia	None.	
Albuminoid Ammonia	None.	

A later analysis of the same water, made by DR. VOELCKER in May, 1898, shows an improvement in the organic constituents.

	Grains per Gallon.	[= Parts per 100,000.]
Total Solid Residue	28·84	41.2
Oxidisable Organic Matter	·11	·16
Nitric Acid	1·87	2·69
Chlorine	2·32	3·31
= Chloride of Sodium	3·82	5·46
Free Ammonia	·003	·004
Albuminoid Ammonia	·003	·004

" The water was colourless and free from deposit."

" The water is a somewhat hard one, owing to the presence of Lime and Magnesia salts, but it contains little dissolved organic matter, and though the amount of Nitrates and chlorides is somewhat high, the water is one which I think may be safely used for drinking purposes."

GORING. (1) From a Well, July, 1894. (2) Another sample from a Well, May, 1894. (3) From a Borehole, April, 1894.

Analysis by R. A. CRIPPS. Communicated by MESSRS. DUKE & OCKENDEN.

	(1)		(2)		(3)	
	Grains per Gallon.	[= Parts per 100,000.]	Grains per Gallon.	[= Parts per 100,000.]	Grains per Gallon.	[= Parts per 100,000.]
Total Solids	48·	[68·]	69·	[98·]	17·	[24·]
Chlorine	9·	[12·8]	13·	[18·]	13·	[18·]
Ammonia	·00588	[·0084]	·0084	[·012]	·00042	[·0006]
Albuminoid Ammonia	·0028	[·004]	·0042	[·006]	·00196	[·0028]
Nitrites	Merest trace.		Small trace.		Absent.	
Nitrates	Trace.		Moderate quantity.		Quantity.	
Temporary Hardness	10°	[14·]	37·3°	[53·]	36°	[51·]
Permanent Hardness	6·5°	[9·]	14°	[20·]	10°	[14·]
Total hardness	16·5°	[23·]	51·3°	[73·]	46°	[65·]

"(1) Although this water contains a large quantity of chlorides these are evidently derived from the soil; they are not accompanied by any excess of nitrates or ammonia; the mere traces of nitrites is probably owing to the well having been recently bored. The water may be safely used for drinking purposes.

(2) Microscopic examination fairly satisfactory. Differs little from a sample examined in April (nitrates somewhat less), presumably No. 3.

(3) Microscopic examination—mineral matters, a few animalcules. This water is of very doubtful purity. Nitrates and chlorides excessive, and these commonly owe their presence to access of sewage, which has become altered in character by the action of the soil."

HASTINGS. Rural District Council.
Analysis by PROF. W. R. SMITH. In grains per gallon.
(1) Silver Hill Well. April, 1895. In Ashdown Sand.
(2) Draper's Well. At the Mill on the higher ground a little south-westward of the above, and about 250 feet above Ordnance Datum. April, 1895. In Tunbridge Wells Sand and Wadhurst Clay.
(3) Experimental Well at Ore, on the northern side of the lane a third of a mile north-east of Christ Church and a little eastward of Windmill. January, 1896. In Ashdown Sand.

	(1)	(2)	(3)
	Grains [=Parts per per Gallon. 100,000.]	Grains [=Parts per per Gallon. 100,000.]	Grains [=Parts per per Gallon. 100,000.]
Colour in 2 feet stratum	Almost colourless	Turbid	Faint blue.
Suspended matter	Very slight	Considerable	Slight, and contained some fibres of clothing.
Taste	Normal	Normal	Normal.
Odour, when heated to 100° F.	Normal	Normal	Normal.
Hardness	17·5° [25·0]	18° [25·7]	5° [7·14]
Total Solid Matter, dried at 120° C.	32· [45·]	29· [41·]	14· [20·]
Loss on Ignition, after recarbonating	10· [14·]	9· [12·8]	5· [7·14]
Total Mineral Matter	22· [31·]	20· [29·]	9· [12·8]
Combined Chlorine. Equal to Common Salt (in 3), 5·8	6· [8·6]	5·8 [8·3]	3·5 [5·]
Nitrogen as Nitrates (no Nitrites)	·5 [·7]	·5 [·7]	·3 [·43]
Ammonia	·004 [·006]	·014 [·02]	·007 [·01]
Albuminoid Ammonia	·003 [·004]	·007 [·01]	·003 [·004]

"(1) The high chlorine is clearly geological. This water may be used with confidence for all domestic purposes.

(2) The suspended matter consists of vegetable *débris* and starchy matters, which ought not to be found in potable waters, and can hardly be due to any cause but contamination with surface-water. [MR. W SKILLER tells us that the cause of this was found out and cut off, after which a futher analysis proved the water to be satisfactory.]

(3) This water is of a high degree of organic purity."

HASTINGS. Dr. Maccabe's Spring. Feb. 21, 1873.
Rivers Pollution Commission, 6th Report, 1874, p. 121.
Temperature, 10·0° C. Clear and palatable.

Parts per 100,000.

Total Solid Impurity	14·92
Organic Carbon	·024
Organic Nitrogen	·005
Ammonia	—
Nitrogen as Nitrates or Nitrites	·433
Total Combined Nitrogen	·438
Chlorine	4·70
Hardness, Temporary	·3
,, Permanent	5·7
,, Total	6·0

ANALYSES.

HENFIELD. From a Borehole. July, 1895.
Analysis by R. A. CRIPPS. Communicated by MESSRS. DUKE & OCKENDEN.

	Grains per Gallon.	[= Parts per 100,000.]
Total Solids	24·	[34·
Chlorine	1·85	2·64
Ammonia	·0098	·014
Albuminoid Ammonia	·000224	·003]
Nitrites	Absent.	
Nitrates	Merest trace.	
Hardness, Temporary	12·05°	[17·21
" Permanent	2·3°	3·29
" Total	14·35°	20·5]

"Microscopic examination satisfactory. Water of good quality. Sample slightly cloudy when received : the water contains a little iron, and this is undoubtedly the cause of the trouble."

HOLLINGTON. Well for Hastings Waterworks. February, 1874.
Analysis by DR. A. VOELCKER.
[Water from the Ashdown Sand].

	Grains per Gallon.	[= Parts in 100,000.]
Solid residue, dried at 140° C., in which (found by direct determination) organic and volatile matter 1·12, including ·224 Oxydisable Organic Matter	13·44	[19·2
Lime	3·05	4·36
Magnesia	·55	·78
Sulphuric Acid	·96	1·37
Chlorine	2·86	4·08
Soluble Silica	·28	·4]
Alkalis and Carbonic Acid, not determined separately.		
Free (saline) Ammonia	·009	[·0128
Organic (albuminoid) Ammonia	·002	·0028]
The components may be represented as follows—		
Organic and Volatile Matter [as above]	1·12	[1·6
Carbonate of Lime	4·25	6·07
Sulphate of Lime	1·63	2·33
Carbonate of Magnesia	1·15	1·64
Chloride of Sodium	4·21	6·01
Alkaline Carbonates	·8	1·14
Soluble Silica	·28	·4
Hardness before boiling	6¾°	9·64
" after boiling	3½°	5]

"The water was clear and colourless. The residue left on evaporation was only slightly coloured yellow by a little vegetable matter. The water contains no nitrates and is free from animal organic impurities. I consider it of first rate quality, wholesome and good for drinking and well suited for cooking and washing."

HORSHAM. London and Brighton Railway Station. Well. 1881.

Analysis by BERNARD DYER.

[Water from the Tunbridge Wells Sand.]

	Grains per Gallon.	[= Parts per 100,000.]
Sulphate of Lime	3·62	5·174
Sulphate of Magnesia	·17	·243
Carbonate of Magnesia	1·12	1·6
Nitrate of Magnesia	·03	·043
Chloride of Sodium	2·54	3·63
Oxide of Iron, &c.	·49	·7
Alkaline Carbonates & Organic Matter	4·91	7·0
Total Solid Matter in solution	12·88	18·4
Phosphoric Acid	Strong traces.	
Free Ammonia	·035	·05
Albuminised Ammonia	·002	·0028
Nitrogen as Nitrates	·006	·0085
Hardness before boiling	4°	5·7
,, after boiling	0¾°	1·07

"As a boiler water, this sample leaves nothing to be desired. It contains less than 13 grains of solid dissolved matter per gallon, of which scarcely 5 grains consist of earthy salts, the remainder being simply common salt and alkaline carbonates—which latter are rather beneficial than otherwise."

HORSTED GREEN, see p. 121.

LEWES. (1) Springs in Verrall's Pool. February 22, 1873. (2) The *Cockshoot Stream* from adjacent Springs. February 22, 1893.

Rivers Pollution Commission, 6th Report, 1874, p. 123.

Water from the [Upper] Chalk. Temperature of (1) 9·0° C., of (2) 8·8° C. 1) Clean and palatable; (2) Slightly turbid; palatable.

	Parts per 100,000.	
	(1)	(2)
Total Solid Impurity	26·44	29·80
Organic Carbon	·057	·087
Organic Nitrogen	·013	·023
Ammonia	·001	·002
Nitrogen as Nitrates or Nitrites	·335	·513
Total Combined Nitrogen	·349	·538
Chlorine	2·30	2·50
Hardness, Temporary	14·2	18·1
,, Permanent	5·1	4·6
,, Total	19·3	22·7

LEWES. Waterworks [Verrall's Pool]. August 10, 1897.

Analysis by JOHN HERON.

	Grains per Gallon.	[= Parts per 100,000.]
Free Ammonia	Trace.	
Albuminoid Ammonia	·006	·0085
Oxygen absorbed in 1 hour	·064	·091
" " 3 hours	·065	·093
Nitrogen as Nitrates	·29	·41
= Nitric Acid	1·30	1·85
Chlorine	1·90	2·71
Total solid matters	20·16	28·8
Hardness before boiling	13°	18·0
" after boiling	3°	4·0

"This sample of water presents a bright clear and sparkling appearance, is perfectly free from sediment and suspended matter. I consider it to be a water of high-class purity and one that may be safely used for drinking and all other domestic purposes."

LITTLEHAMPTON. Anchor Brewery. November, 1869.

Analysis by PROF. W. A. MILLER. Communicated by MR. W. SHELFORD.

[Water from the Upper Chalk.]

	Grains per Gallon.	[= Parts per 100,000.]
Fixed Salts	76·65	109·5
Volatile and Combustible Matters	3·35	4·78
Total Soluble Solids	80·	114·28
Nitric Acid, N_2O_5	1·98	2·83
Ammonia as Salts	·001	·0014
Ammonia from Organic Matter	·008	·0114
Oxygen required to Oxidise Organic Matter by Permanganate	·067	·0957

Appearance clear and brilliant.

Hardness on Clarke's scale 36·9° [52·71].

After boiling an hour 14·9° [21·28].

The water is probably excellent for beer-making, owing to its sulphate of lime, but it is not good for domestic uses.

[The permanent hardness is exceptionally high for a water from the Upper Chalk. The analysis given above does not show that this hardness is due to sulphate of lime. See note on the Well at p. 62.]

Mid Sussex Water Company, see Balcombe.

Newhaven and Seaford. Waterworks. New Well at Poverty Bottom, Denton, 21st April, 1898.

Analysis by O. Hehner.

[Water from the Upper Chalk.]

Parts per 100,000.

Chlorine	3·35
Sulphuric Acid	·34
Nitric Acid	1·30
Phosphoric Acid	None.
Free Ammonia	·0008
Albuminoid Ammonia	·0038
Oxygen absorbed from Permanganate in 15 minutes	·0164
,, ,, in 4 hours	·0240
(both at 80° F.)	
Total Solids	29·96
Loss on ignition	2·04

The composition of the mineral matter was as under :—

Chlorine	3·35
Sulphuric Acid	·34
Nitric Acid	1·30
Silica	·63
Oxide of Iron and Alumina	·23
Lime	11·09
Magnesia	·55
Soda	2·43
Combined Carbonic Acid	8·25
	28·17
Subtract Oxygen for Chlorine	·75
Total mineral matters	27·42

As far as could be ascertained these mineral matters were present in the water in the following forms of combination :—

Sodium Chloride	4·58
Calcium Chloride	·89
Calcium Sulphate	·58
Calcium Nitrate	1·97
Calcium Carbonate	17·3
Magnesium Carbonate	1·1
Silica	·63
Oxide of Iron and Alumina	·23
Total	27·42

"Organically the water is of great purity; there is no evidence of pollution. The character of the water is that of a typical supply from the Chalk. Its hardness is 21·2, 17·4 of which is due to dissolved calcium carbonate. From the analysis alone I say, without hesitation, that the supply is admirably adapted for public use. With the exception of the hardness, which is the normal hardness of pure Chalk water, the supply is faultless."

NEWHAVEN and SEAFORD. Waterworks. Old well.
Water supplied to South Heighton, March, 1897. [From the Upper Chalk.]
Analysis by DR. J. A. VOELCKER. Communicated by MR. T. W. PICKARD.

	Grains per Gallon.	[=Parts per 100,000.]
Total Solid Residue	26·32	37·6
Oxygen absorbed	·020	·029
Lime	8·96	12·8
Magnesia	1·01	1·44
Sulphuric Acid	·77	1·1
Nitric Acid	1·36	1·94
Chlorine	5·51	7·87
= Chloride of Sodium	9·08	12·99
Free Ammonia	·002	·003
Albuminoid Ammonia	·001	·0014

"Colourless but had a little deposit of a blackish colour. It is a somewhat hard water, containing carbonate of lime principally, with some amount of magnesia salts. Chlorides, probably as common salt, are present in considerable quantity. The water contains very little Ammonia and has no excess of organic matter in solution. Nitrates are present to some extent but are hardly excessive, and though chlorides exist in certainly large amount, these may arise from natural sources, and I am not inclined to attribute them to pollution. The water, though, in my opinion, not one that can be called a thoroughly good or high-class one, can, I think, be considered a fit one for drinking purposes."

RUSTINGTON. (1) From a Borehole, August, 1894. (2) From a Borehole, May, 1895. (3) From a Well. (4) From a Borehole.
Analysis by R. A. CRIPPS. Communicated by MESSRS. DUKE & OCKENDEN.

	In Grains per Gallon [=Parts per 100,000.]			
	(1)	(2)	(3)	(4)
Total Solids	40·5 [58·]	37· [53·]	48· [68·]	71·5 [102·14]
Chlorine	7·15 [10·21]	4·8 [6·85]	6·05 [8·64]	11·45 [16·35]
Ammonia	·0021 [·003]	merest trace.	small trace.	trace.
Albuminoid Ammonia	·00112 [·0016]	·0021 [·003]	·0021 [·003]	·0021 [·003]
Nitrites absent, Nitrates	trace.	small quantity.	moderate q'ntity	large quantity.
Temporary Hardness	—	15·70 [22·4]	39·5° [56·4]	35° [50]
Permanent "	—	4·3° [6·1]	8·4° [12·0]	7·6° [10·8]
Total "	22·2° [31·7]	20° [28·5]	47·9° [68·4]	42·6° [60·8]

"(1) Microscopic examination, mineral matter. This water is of good quality.
(2) Microscopic examination satisfactory. This water is of good quality. The quantity of chlorine is probably explained by the proximity of the sea.
(3) Microscopic examination satisfactory. This water is of fair quality; but contains rather large quantities of chlorides and nitrates. It may be used for drinking purposes, although it cannot be classed as first-class.
(4) This water is of very doubtful purity. The Nitrates and Chlorides are excessive, and these probably owe their presence to pollution with sewage, which has become altered in character by the action of the soil."

RYE. Sample from Rye Hill Reservoir. The supply is from springs at the base of the old cliff, near by, just within the borough boundary.

Analysis by PROF. J. ATTFIELD, November 1894.

	Grains per Gallon.	[=Parts per 100,000.]
Suspended Solid Matter, dried at 250° F.	None.	
Dissolved ,, ,, ,,	25·	[35·7]
Ammoniacal Matter, yielding 10 per cent. of Nitrogen (= Ammonia per million ·08)	·046	[·0657]
Albuminoid Organic Matter, yielding 10 per cent. of Nitrogen (= Ammonia per million ·02)	·012	[·0171]
Nitrates (no Nitrites), containing 17 per cent. of Nitrogen (= Nitrogen ·58)	3·48	[4·97]
Chlorides, containing 60 per cent. of Chlorine (= Chlorine 4·4)	7·3	[10·43]
Oxygen absorbed in three hours	0·2	[·028]
Hardness, removed by boiling, 7·5° [10·71]	12·5°	[17·85]
Hardness, unaffected by boiling, 5·0° [7·14]		

Water clear and bright. Of excellent quality.

The water from the proposed site for further supply, at the foot of the old cliff, about a quarter of a mile N.E. of Cadborough, gave a like analysis.

ST. LEONARDS-ON-SEA. Spring in Railway Tunnel. Feb. 21st, 1873.

Rivers Pollution Commission, 6th Report, 1874, p. 127.

Water from the Hastings Sand. Temperature, 4·0° C.

Turbid. Palatable.

	Parts per 100,000.
Total Solid Impurity	41·92
Organic Carbon	·224
Organic Nitrogen	·054
Ammonia	·088
Nitrogen as Nitrates or Nitrites	·478
Total Combined Nitrogen	·604
Chlorine	9·60
Hardness, Temporary	4·0
,, Permanent	12·9
,, Total	16·9

SEAFORD, see NEWHAVEN AND SEAFORD pp. 116, 117.

SOUTH HEIGHTON. Sussex Portland Cement Works. March 1897.

Analysis by DR. J. A. VOELCKER. Communicated by MR. T. W. PICKARD,

[Water from the Upper Chalk.]

	Grains per Gallon.	[= Parts per 100,000.]
Total Solid Residue	24·39	34·8
Oxygen absorbed	·027	·039
Lime	7·84	11·2
Magnesia	·70	1·
Sulphuric Acid	·96	1·37
Nitric Acid	2·73	3·9
Chlorine	3·39	4·84
= Chloride of Sodium	5·59	7·98
Free Ammonia	·006	·0085
Albuminoid Ammonia	trace.	

"This water was colourless but contained some white flocculent deposit. . . . . It does not contain any quantity of dissolved organic matter, but there is more ammonia than in either 1 or 2 [Seaford Waterworks or Glynde Butter Factory]. Chlorides, again, are in excess of those in 2, and there is even more nitric acid (as nitrates). This latter feature, as in the case of water 2, indicates, in my opinion, the existence of pollution of the supply, and for that reason I do not regard the source as a satisfactory one." [The well is close to the marshes of the Ouse.]

TELSCOMBE. Warren Farm. Brighton Industrial School.

Rivers Pollution Commission, 6th Report, 1874, p. 97.

Water from [the Lower Greensand at] 1285 feet. Temperature 9·9° C.
Water clear and palatable.

	Parts per 100,000.
Total Solid Impurity	35·36
Organic Carbon	·078
Organic Nitrogen	·007
Ammonia	—
Nitrogen as Nitrates and Nitrites	·068
Total Combined Nitrogen	·075
Chlorine	8·40
Hardness, Temporary	3·2
„ Permanent	1·2
„ Total	4·4

[The temperature (9·9° C.) is apparently that of the water standing in the well, not that of the spring 1,000 feet lower. The latter is inaccessible, the lower shaft not being vertically under the upper well, which contains a hundred feet of water, see p. 83.]

WASHINGTON. From a Borehole opposite the Church.

Analysis by R. A. CRIPPS. Communicated by MESSRS. DUKE & OCKENDEN.

	Grains per gallon.	[= Parts per 100,000.]
Total Solids	34·	[48·5]
Chlorine	2·6	[3·71]
Ammonia	·00308	[·0044]
Albuminoid Ammonia	·0028	[·004]
Nitrites	absent	
Nitrates	traces	
Hardness Temporary	16·35°	[23·35]
" Permanent	6·4°	[9·15]
" Total	22·75°	[32·5]

"Water of excellent quality for drinking-purposes, and free from organic pollution."

WORTHING. New Well at Waterworks. July 17th, 1868.

Rivers Pollution Commission, 6th Report, 1874, p. 99.

Water from the Upper Chalk. Clear and palatable.

	Parts per 100,000.
Total Solid Impurity	32·44
Organic Carbon	·007
Organic Nitrogen	—
Ammonia	·002
Nitrogen as Nitrates or Nitrites	·420
Total Combined Nitrogen	·422
Chlorine	3·08
Hardness Temporary	16·4
" Permanent	8·3
" Total	24·7

WORTHING. From a bored Well.

Analysis by R. A. CRIPPS. Communicated by MESSRS. DUKE & OCKENDEN.

	Grains per Gallon.	[= Parts per 100,000.]
Total Solids	23·5	[33·6]
Chlorine	1·9	[2·7]
Ammonia	·00112	[·0016]
Albuminoid Ammonia	·00084	[·0012]
Nitrites	merest trace	
Nitrates	small	

"Water of good quality. It is exceptionally free from organic matter and the amount of dissolved saline substances is moderate. Microscopic examination satisfactory."

POSTSCRIPT.

Haywards Heath. County Lunatic Asylum.

Analysis by Dr. Letheby.

[Water from the **Tunbridge Wells Sand**].

	Grains per Gallon.	[= parts per 100,000].
Carbonate of Lime	10·01	14·3
Carbonate of Magnesia	·72	1·03
Sulphate of Lime	1·68	2·4
Sulphate of Magnesia	·61	·84
Chloride of Sodium	3·16	4·5
Silica, Alumina and **Peroxide of Iron**	1·15	1·64
Organic Matter	·06	·09
Hardness Total	11°	15·5
Hardness Permanent	3°	4·0

Dr. C. E. Saunders adds (1890) that "the oxidation of the iron contained in solution was a source of much trouble in the early days, for it choked the pipes, and discoloured the clothing. This has to some extent been obviated by pumping the water through a fountain, whereby it becomes aërated and deposits the iron peroxide."

Horsted Green. Stroodland Farm. 1898.

Analysis by Mr. S. A. Woodhead, communicated by Mr. Charles Dawson

	Grains per Gallon	[= parts per 100,000].	
Total solids	100·1	143·	
Solids after ignition	78·1	111·6	
Chlorine	2·6	3·7	
Ammonia (free)	·074	} parts per million.	
Ammonia (Albuminoid)	·062		
Oxygen taken from permanganate in ¼ hour	none.		
" " " " 4 hours	trace.		
Nitrogen as Nitrates and Nitrites	·043	·061	
Nitrites	absent.		
Hardness (total)	52°	74·	
Hardness after boiling	40°	59·	
Phosphates	absent.		
Metallic impurity	**none.**		

"**The water was** clear and palatable, **and in** warming no disagreeable odour **was noticed**. On standing there was practically no sediment. On account **of the** excessive hardness it cannot be recommended for boiler use, nor yet **for** drinking purposes owing **to the solids** in solution, consisting largely **of** magnesium salts."

INDEX.

Aldingbourne, 8.
Aldrington waterworks, 74.
Ambergate, 38.
Amberstone, 38.
Analyses, 103.
Angmering, 8, 103.
Arundel, 8, 103, 104.
Ashburnham Place, 8.
Ashdown Sand, 3.
Ashfield, 64.
Bagshot Beds, 5.
Balcombe, 9, 104.
Barcombe, 9.
Barnham, 10.
————— Junction, 27.
Battle, 10.
————, Sub-wealden Exploration, 65.
Beddingham, 11, 105.
Beeding, Lower, 11.
————— Upper, 85.
Beedingwood, 11.
Bexhill, 12, 13, 14, 101.
Birdham, 15.
Blatchington, East, 24.
Blue-coat School, 53.
Bognor, 15.
————— waterworks, 28, 64.
Bopeep, 47.
Borough Farm, 75.
Bosham, 15.
Bracklesham Beds, 5.
Brede Valley, 91–96.
Brighton, 16.
————— Industrial School, 83, 119.
————— waterworks, 16, 17, 72, 105, 106.
Broadwater, 17, 99, 106.
Broughton Spring, 56, 57.
Buckhurst Manor Farm, 86.
Buckshole reservoir, 49.
Bulverhithe, 42.
Burpham, 106.
Buxted, 17.
Cadborough Cliff, 78.
Catsfield, 18.
Chalk, 5.
Chichester, 18, 19, 107.
Chiddingly, 20.
Chiltington, 20.
Christ's Hospital, 53.
Claypits Farm, 33.
Coal, Trial for, 14.
Cobb's Nest, 79.
Cooksbridge 20.
Corseley Farm, 38.
County Lunatic Asylum, 45, 121.
Courthouse Farm, 11, 105.
Crowham, 92.
Crawley, 20, 107.
Crowborough, 77.
————— Warren, 102.
Crowhurst, 21.
Cuckfield, 22.
————— Place, 23.

Dalepark, 62.
Dean, West, 90, 91.
Ditchling, 23.
Drift deposits, 6.
Easebourn, 23.
Eason's Green, 33.
East Blatchington, 24.
Eastbourne, 24–27, 108–109.
————— waterworks, 26, 27, 33, 36, 56, 57, 90, 91, 97, 103, 109.
Eastergate, 27, 28.
East Grinstead, 28, 29, 33, 110.
East Hoathly, 33.
Ecclesbourne valley, 29, 30.
Elstead, 29.
Eridge Park, 36.
Fairlight, 29, 30.
————— Clay, 3.
Fay Gate, 31.
Filsham, 47–50.
Fishbourne, 31, 32.
Fittleworth, 32.
————— Scrub House, 58.
Folkington, 33, 56.
Forest Row, 33.
Forge Bridge, 96.
Formations, List of, 1.
Framfield, 33.
Frant, 33–36.
Friston, 36, 109.
Funtington, 37.
Gault, 4.
Geological Survey publications, 7.
Gilders' Oak Farm, 64.
Glynde, 37, 110, 111.
Goldstone Bottom, 16, 105.
Goring, 111.
Graylings Well Farm, 18.
Greensands, 4.
Grinstead, East, 28, 29, 33, 110.
Groombridge, 38.
Gullands Oak, 64.
Hailsham, 38, 39.
Hambrook House, 37.
Hartfield 39, 40.
Hartwell 39, 40.
Hassocks, 57.
Hastings, 41–44, 112.
————— waterworks, 18, 21, 29, 30, 42–44, 47–52, 91–96, 113.
Haywards Heath, 44, 45, 121.
Headhone Farm, 8.
Heathfield, 45, 87–89.
Heighton, South, 119.
Hellingly, 46.
Hempstead Mill, 85.
Henfield, 46, 113.
Hermongers, 77.
Hillhouse Farm, 78.
Hoathley, East, 33.
Hollington, 47–52, 113.
Holywell Springs, 108, 109.
Horsham, 11, 53, 54, 114.

Horsted Green, 121.
—— Keynes, 55.
——, Little, 62.
Hunston, 56.
Jevington, 56, 57.
Jurassic rocks, 3.
Keymer, 57, 58.
Kingsfold estate, 90.
Kingston, 58.
Kirdford, 58.
Knowle, 33.
Lamberhurst, 59.
Lancing, 59.
Langley Farm, 97.
Laughton, 59.
Leylands Park, 58.
Lewes, 60, 114, 115.
Littlehampton, 61, 62, 115.
Little Horsted, 62.
Lodsworth, 62.
Lower Beeding, 11.
Lower Greensand, **4**.
Loxwood, 62.
Madehurst, 62.
Manning's Heath, **71**.
Mayfield, 63.
Maynard's Gate, **77**.
Merston, 64.
Merrywood, **84**.
Midhurst, 23, **64, 65**.
Mid-Sussex Waterworks, **9**, 104.
Moor Lane, 75.
Mountfield, 65–70.
Mundham, North, 71.
Netherfield, 65–70.
Newhaven and Seaford Waterworks, 24, 116, 117.
Newick, 70.
Newmarket, Kingston, **58**.
New Timber, 71.
Norlington Green, **75**.
Normanhurst Court, **18**.
North Mundham, 71.
Nuthurst, 71.
Offham Farm, 104.
Old Roar, 49–52.
Ore, 112.
Owl's Castle, 92.
Pagham, 72.
Palæozoic rocks, **2**.
Park Croft, **98**.
—— Farm, Hellingley, 46.
—— —— Selsey, 79.
—— House, 75.
Patcham, 72.
Patching, 73.
Pepsham (or Pepplesham), **42**.
Petworth House, 73.
Pevensey Sluice, 73.
Plashett Park, 73.
Polegate, 39.
Portslade, 74.
Pulborough, **32, 75**.
Rainfall, **6**.
Reading Beds, **5**.
Redley Farm, 93.

Ringmer, 75.
Room Bottom, 85.
Rotherfield, 77.
Rowfold Farm, 81.
Rudgwick, 77, 78.
Rustington, 117.
Rye, **78, 118**.
St. Leonards-on-Sea, 118.
—— —— Waterworks, 41.
St. Mary Bulverhithe, 42.
Seaford Waterworks, 24, 116, 117.
Sefter School, **72**.
Selmeston, 79.
Selsey, 79, 80.
Shipley, 80, 81.
Short Gate, **81**.
Silver Hill, 52, 112.
Slaugham, 81.
Slinfold, 81, 82.
Sompting, 82.
South Heighton, 119.
Stammerham, 53.
Steyning, 82.
—— Water Supply, **85**.
Stone Cross, 20.
—— Lodge, 11.
Stroodland Farm, 121.
Sub-Wealden Exploration, 65–70.
Sussex Portland Cement Works, 119.
Swanbourne, 103.
Telscombe, 83, 119.
Thakeham, 84.
Three Bridges Station, 99.
Ticehurst, 84.
Timber, New, 71.
Tunbridge Wells Sand, 3.
—— —— —— Station, 34, 35.
Uckfield, **84, 85**.
Upper Beeding, **85**.
Upper Greensand, **4**.
Verrall's Pool, 60, 114, 115.
Wadhurst, 86.
—— Clay, 3.
Walberton, 86.
Waldron, 87–89.
Wannock, 33, 56, 57, 109.
Warnham, 89, 90.
Warren Farm, 83, 119.
Washington, 120.
Water Analyses, 103–121.
—— character of, in different strata, 1, 2.
Weald Clay, 4.
West Dean, 90, 91.
Westfield, 91–96.
West Firle, 97.
Westham, 97, 98.
Willingdon, 98.
Willowhurst, 20.
Winchelsea, 98.
Withyham, 102.
Wivelsfield, 98.
Woolwich and Reading Beds, 5.
Worth, 99.
Worthing, 99–101, 120.

www.ingramcontent.com/pod-product-compliance
Lightning Source LLC
Chambersburg PA
CBHW020121170426
43199CB00009B/586